Mechanical Insulating

Level One

Trainee Guide

P Pearson

Boston   Columbus   Indianapolis   New York   San Francisco   Amsterdam
Cape Town Dubai London Madrid Milan Munich Paris Montreal Toronto Delhi
Mexico City Sao Paulo Sydney Hong Kong Seoul Singapore Taipei Tokyo

NCCER

President: Don Whyte
Vice President: Steve Greene
Chief Operations Officer: Katrina Kersch
Mechanical Insulating Project Managers: Elizabeth Schlaupitz,
  Jamie Carroll
Senior Development Manager: Mark Thomas

Senior Production Manager: Tim Davis
Quality Assurance Coordinator: Karyn Payne
Desktop Publishing Coordinator: James McKay
Permissions Specialists: Kelly Sadler, Adrienne Payne
Production Specialist: Kelly Sadler
Editors: Graham Hack

Writing and development services provided by Topaz Publications, Liverpool, NY

Lead Writer/Project Manager: Thomas Burke
Desktop Publisher: Joanne Hart
Art Director: Alison Richmond

Permissions Editor: Andrea LaBarge
Writers: Troy Staton, Thomas Burke, Terry Egolf, Veronica
  Westfall, Carol Hebert, Darrell Wilkerson

Pearson
Director of Alliance/Partnership Management: Andrew Taylor
Editorial Assistant: Collin Lamothe
Program Manager: Alexandrina B. Wolf
Director of Marketing: Leigh Ann Simms

Senior Marketing Manager: Brian Hoehl
Composition: NCCER

Text Fonts: Palatino and Univers

ISBN-13:   978-0-13-413099-6
ISBN-10:   0-13-413099-5

Preface

To the Trainee

Insulation is a crucial element for the operation of mechanical systems and provides thermal protection for boilers, HVAC systems, duct work, and piping. Insulating these systems improves operating efficiency by managing system temperature, controlling and preventing the presence of condensation and mold, and protecting personnel from hot or cold surfaces.

Craft professionals in the mechanical insulating industry must learn both the fundamentals of the industry as well as specific skills and knowledge for their craft. Many are trained through craft apprenticeship programs, which provide both classroom instruction and on-the-job learning.

Workers in the industry are well paid and, according to the Bureau of Labor Statistics, constitute a workforce of over 25,000 workers. Regardless of where your career may take you, NCCER's Mechanical Insulating curriculum will provide you a solid foundation on your road to success.

New with *Mechanical Insulating Level One*

NCCER is pleased to release *Mechanical Insulating* in full color, with new photographs and figures. This edition is now presented in NCCER's improved instructional systems design, in which the sections of each module are directly tied to learning objectives.

The curriculum has also been thoroughly redesigned, updated, and expanded. Level One has been reorganized to provide an overview of the fundamentals of mechanical insulation by incorporating modules previously found in subsequent levels. Some modules were removed due to their content being incorporated into existing modules.

We wish you success as you progress through this training program. If you have any comments on how NCCER might improve upon this textbook, please complete the User Update form located at the back of each module and send it to us. We will always consider and respond to input from our customers.

We invite you to visit the NCCER website at **www.nccer.org** for information on the latest product releases and training, as well as online versions of the *Cornerstone* magazine and Pearson's NCCER product catalog.

Your feedback is welcome. You may email your comments to **curriculum@nccer.org** or send general comments and inquiries to **info@nccer.org**.

NCCER Standardized Curricula

NCCER is a not-for-profit 501(c)(3) education foundation established in 1996 by the world's largest and most progressive construction companies and national construction associations. It was founded to address the severe workforce shortage facing the industry and to develop a standardized training process and curricula. Today, NCCER is supported by hundreds of leading construction and maintenance companies, manufacturers, and national associations. The NCCER Standardized Curricula was developed by NCCER in partnership with Pearson, the world's largest educational publisher.

Some features of the NCCER Standardized Curricula are as follows:

- An industry-proven record of success
- Curricula developed by the industry, for the industry
- National standardization providing portability of learned job skills and educational credits
- Compliance with the Office of Apprenticeship requirements for related classroom training (*CFR 29:29*)
- Well-illustrated, up-to-date, and practical information

NCCER also maintains the NCCER Registry, which provides transcripts, certificates, and wallet cards to individuals who have successfully completed a level of training within a craft in NCCER's Curricula. *Training programs must be delivered by an NCCER Accredited Training Sponsor in order to receive these credentials.*

Special Features

In an effort to provide a comprehensive and user-friendly training resource, this curriculum showcases several informative features. Whether you are a visual or hands-on learner, these features are intended to enhance your knowledge of the construction industry as you progress in your training. Some of the features you may find in the curriculum are explained below.

Introduction

This introductory page, found at the beginning of each module, lists the module Objectives, Performance Tasks, and Trade Terms. The Objectives list the knowledge you will acquire after successfully completing the module. The Performance Tasks give you an opportunity to apply your knowledge to real-world tasks. The Trade Terms are industry-specific vocabulary that you will learn as you study this module.

Trade Features

Trade features present technical tips and professional practices based on real-life scenarios similar to those you might encounter on the job site.

Bowline Trivia

Some people use this saying to help them remember how to tie a bowline: "The rabbit comes out of his hole, around a tree, and back into the hole."

Figures and Tables

Photographs, drawings, diagrams, and tables are used throughout each module to illustrate important concepts and provide clarity for complex instructions. Text references to figures and tables are emphasized with *italic* type.

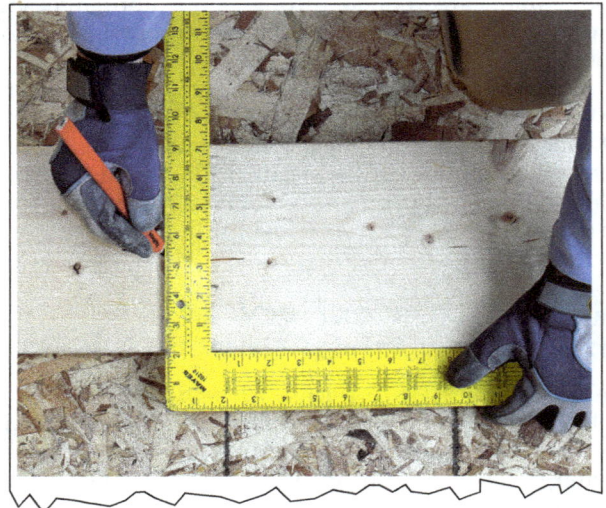

Notes, Cautions, and Warnings

Safety features are set off from the main text in highlighted boxes and categorized according to the potential danger involved. Notes simply provide additional information. Cautions flag a hazardous issue that could cause damage to materials or equipment. Warnings stress a potentially dangerous situation that could result in injury or death to workers.

> **NOTE**
> Nameplates must be posted on each material handling device. The nameplate must indicate

> **CAUTION**
> It is essential to note the revision designation on a construction drawing and to use only the latest

> **WARNING!**
> Saw teeth are very sharp. Use gloves and do not handle the saw teeth with bare hands. When cutting with a saw, ensure that your fingers remain clear of the teeth at all times.

Case History

Case History features emphasize the importance of safety by citing examples of the costly (and often devastating) consequences of ignoring best practices or OSHA regulations.

> **Case History**
>
> **Requesting an Outage**
>
> An electrical contractor requested an outage when asked to install two bolt-in, 240V breakers in panels in a data processing room. It was denied due to the 24/7 worldwide information processing hosted by the facility. The contractor agreed to proceed only if the client would sign a letter agreeing not to hold them responsible if an event occurred that damaged computers or resulted in loss of data. No member of upper management would accept liability for this possibility, and the outage was scheduled.
>
> **The Bottom Line:** If you can communicate the liability associated with an electrical event, you can influence management's decision to work energized.

Going Green

Going Green features present steps being taken within the construction industry to protect the environment and save energy, emphasizing choices that can be made on the job to preserve the health of the planet.

> **GOING GREEN**
>
> **Reducing Your Carbon Footprint**
>
> Many companies are taking part in the paperless movement. They reduce their environmental impact by reducing the amount of paper they use. Using email helps to reduce the amount of paper used,

Did You Know

Did You Know features introduce historical tidbits or interesting and sometimes surprising facts about the trade.

> **Did You Know?**
>
> **Safety First**
>
> Safety training is required for all activities. Never operate tools, machinery, or equipment without prior training. Always refer to the manufacturer's instructions.

Step-by-Step Instructions

Step-by-step instructions are used throughout to guide you through technical procedures and tasks from start to finish. These steps show you how to perform a task safely and efficiently.

> Perform the following steps to erect this system area scaffold:
>
> *Step 1* Gather and inspect all scaffold equipment for the scaffold arrangement.
>
> *Step 2* Place appropriate mudsills in their approximate locations.
>
> *Step 3* Attach the screw jacks to the mudsills.

Trade Terms

Each module presents a list of Trade Terms that are discussed within the text and defined in the Glossary at the end of the module. These terms are presented in the text with bold, blue type upon their first occurrence. To make searches for key information easier, a comprehensive Glossary of Trade Terms from all modules is located at the back of this book.

> During a rigging operation, the load being lifted or moved must be connected to the apparatus, such as a crane, that will provide the power for movement. The connector—the link between the load and the apparatus—is often a sling made of synthetic, chain, or wire rope materials. This section focuses on three types of slings:

Section Review

Each section of the module wraps up with a list of Additional Resources for further study and Section Review questions designed to test your knowledge of the Objectives for that section.

> **Additional Resources**
>
> *Materials Handling Handbook*, The American Society of Mechanical Engineers (ASME) and The International Material Management Society (IMMS), Raymond A. Kulwiec, Editor-in-Chief. 1985. New York, NY: Wiley-Interscience.
>
> *Manufacturing Facilities Design & Material Handling*, Matthew P. Stevens, Fred E. Meyers. 2013. West Lafayette, IN: Purdue University Press.
>
> **1.0.0 Section Review**
>
> 1. For material handling tasks, it is just as important to be mentally fit as it is to be ____.
>   a. physically fit
>   b. physically aggressive
>   c. closely supervised
>   d. over 200 pounds
>
> 2. Which of the following is a type of knot that is often used to join the ends of two ropes in non-critical, low-strain applications?
>   a. Bowline
>   b. Clove hitch
>   c. Half hitch
>   d. Square knot

Review Questions

The end-of-module Review Questions can be used to measure and reinforce your knowledge of the module's content.

> **Review Questions**
>
> 1. Identification tags for slings must include the ____.
>   a. type of protective pads to use
>   b. type of damage sustained during use
>   c. color of the tattle-tail
>   d. manufacturer's name or trademark
>
> 2. The type of wire rope core that is susceptible to heat damage at relatively low temperatures is the ____.
>   a. fiber core
>   b. strand core
>   c. independent wire rope core
>   d. metallic link supporting core
>
> 3. Synthetic slings must be inspected ____.
>   a. once every month
>   b. visually at the start of each work week
>   c. before every use
>   d. once wear or damage becomes apparent
>
> 4. An alloy steel chain sling must be removed from service if there is evidence that ____.
>   a. the sling has been used in different hitch configurations
>   b. replacement links have been used to repair the chain
>   c. the sling has been used for more than one year
>   d. strands in the supporting core have weakened
>
> 5. A piece of rigging hardware used to couple the end of a wire rope to eye fittings, hooks, or other connections is a(n) ____.
>   a. eyebolt
>   b. hitch
>   c. shackle
>   d. U-bolt
>
> 6. A lifting clamp is most likely to be used to move loads such as ____.
>   a. steel plates
>   b. piping bundles
>   c. concrete blocks
>   d. plastic tubing
>
> 7. Chain hoists are able to lift heavy loads by utilizing a ____.
>   a. rope and pulley system
>   b. rigger's strength
>   c. stationary counterweight
>   d. gear system
>
> 8. Before attempting to lift a load with a chain hoist, make sure that the ____.
>   a. hoist is secured to a come-along
>   b. load is properly balanced
>   c. tag lines are properly anchored
>   d. tackle is connected to its power source
>
> 9. A hitch configuration that allows slings to be connected to the same load without using a spreader beam is a ____.
>   a. double-wrap hitch
>   b. choker hitch
>   c. bridle hitch
>   d. basket hitch
>
> 10. To make the emergency stop signal that is used by riggers, extend both arms ____.
>   a. horizontally with palms down and quickly move both arms back and forth
>   b. directly in front and then move both arms up and down repeatedly
>   c. vertically above the head and wave both arms back and forth
>   d. horizontally with clenched fists and move both arms up and down
>
> 00108   Introduction to Basic Rigging   Module Six 31

NCCER Standardized Curricula

NCCER's training programs comprise more than 80 construction, maintenance, pipeline, and utility areas and include skills assessments, safety training, and management education.

Boilermaking
Cabinetmaking
Carpentry
Concrete Finishing
Construction Craft Laborer
Construction Technology
Core Curriculum: Introductory
 Craft Skills
Drywall
Electrical
Electronic Systems Technician
Heating, Ventilating, and Air
 Conditioning
Heavy Equipment Operations
Heavy Highway Construction
Hydroblasting
Industrial Coating and Lining
 Application Specialist
Industrial Maintenance Electrical
 and Instrumentation Technician
Industrial Maintenance Mechanic
Instrumentation
Ironworking
Masonry
Mechanical Insulating
Millwright
Mobile Crane Operations
Painting
Painting, Industrial
Pipefitting
Pipelayer
Plumbing
Reinforcing Ironwork
Rigging
Scaffolding
Sheet Metal
Signal Person
Site Layout
Sprinkler Fitting
Tower Crane Operator
Welding

Maritime

Maritime Industry Fundamentals
Maritime Pipefitting
Maritime Structural Fitter

Green/Sustainable Construction

Building Auditor
Fundamentals of Weatherization
Introduction to Weatherization
Sustainable Construction
 Supervisor
Weatherization Crew Chief
Weatherization Technician
Your Role in the Green
 Environment

Energy

Alternative Energy
Introduction to the Power Industry
Introduction to Solar Photovoltaics
Power Generation Maintenance
 Electrician
Power Generation I&C
 Maintenance Technician
Power Generation Maintenance
 Mechanic
Power Line Worker
Power Line Worker: Distribution
Power Line Worker: Substation
Power Line Worker: Transmission
Solar Photovoltaic Systems Installer
Wind Energy
Wind Turbine Maintenance
 Technician

Pipeline

Abnormal Operating Conditions,
 Control Center
Abnormal Operating Conditions,
 Field and Gas
Corrosion Control
Electrical and Instrumentation
Field and Control Center
 Operations
Introduction to the Pipeline
 Industry
Maintenance
Mechanical

Safety

Field Safety
Safety Orientation
Safety Technology

Supplemental Titles

Applied Construction Math
Tools for Success

Management

Construction Workforce
 Development Professional
Fundamentals of Crew Leadership
Mentoring for Craft Professionals
Project Management
Project Supervision

Spanish Titles

Acabado de concreto: nivel uno
 (*Concrete Finishing Level One*)
Aislamiento: nivel uno
 (*Insulating Level One*)
Albañilería: nivel uno
 (*Masonry Level One*)
Andamios (*Scaffolding*)
Carpintería: Formas para
 carpintería, nivel tres
 (*Carpentry: Carpentry Forms, Level
 Three*)
Currículo básico: habilidades
 introductorias del oficio
 (*Core Curriculum: Introductory Craft
 Skills*)
Electricidad: nivel uno
 (*Electrical Level One*)
Herrería: nivel uno
 (*Ironworking Level One*)
Herrería de refuerzo: nivel uno
 (*Reinforcing Ironwork Level One*)
Instalación de rociadores: nivel uno
 (*Sprinkler Fitting Level One*)
Instalación de tuberías: nivel uno
 (*Pipefitting Level One*)
Instrumentación: nivel uno, nivel
 dos, nivel tres, nivel cuatro
 (*Instrumentation Levels One through
 Four*)
Orientación de seguridad
 (*Safety Orientation*)
Paneles de yeso: nivel uno
 (*Drywall Level One*)
Seguridad de campo
 (*Field Safety*)

Acknowledgments

This curriculum was revised as a result of the farsightedness and leadership of the following sponsors:

Brace Industrial Group
Brock Services LLC
Industrial Construction & Engineering Co.
Insulation Specialties Inc.
L&C Insulation

Marquis Construction Services, Inc.
National Insulation Association
Petrin Corporation
Southern Tier Insulations

This curriculum would not exist were it not for the dedication and unselfish energy of those volunteers who served on the Authoring Team. A sincere thanks is extended to the following:

Randy Beard
Tim Coughlin
Jeffrey DeGraaf
Peter Gauchel
David Gottlich

Wes Howard
Michele Jones
Ron King
Al Olvera
Gary Satterfield

NCCER Partners

American Council for Construction Education
American Fire Sprinkler Association
Associated Builders and Contractors, Inc.
Associated General Contractors of America
Association for Career and Technical Education
Association for Skilled and Technical Sciences
Construction Industry Institute
Construction Users Roundtable
Design Build Institute of America
GSSC – Gulf States Shipbuilders Consortium
ISN
Manufacturing Institute
Mason Contractors Association of America
Merit Contractors Association of Canada
NACE International
National Association of Women in Construction
National Insulation Association
National Technical Honor Society
National Utility Contractors Association
NAWIC Education Foundation
North American Crane Bureau
North American Technician Excellence
Pearson

Prov
SkillsUSA®
Steel Erectors Association of America
U.S. Army Corps of Engineers
University of Florida, M. E. Rinker Sr., School of Construction Management
Women Construction Owners & Executives, USA

NCCER Business Partners

Contents

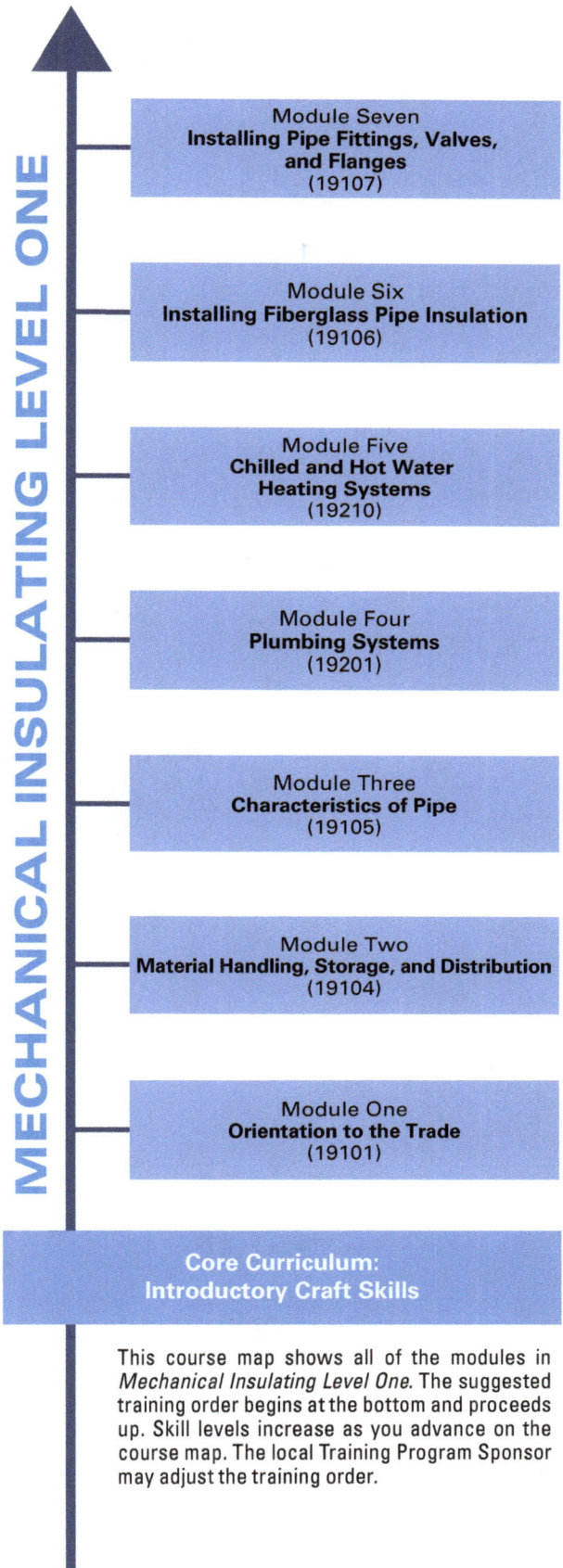

Module Seven

Insulating Pipe Fittings, Valves, and Flanges

Identifies and presents the insulation requirements for different types of valves, fittings, and joints found in commercial and industrial pipe systems. This module describes insulation practices related to non-flanged and flanged valves, non-flanged fittings and joints, flange fittings, and mechanical fittings. It also explains calculations used when insulating ells. (Module ID 19107; 40 Hours)

Glossary

Index

MECHANICAL INSULATING LEVEL ONE

Module Seven
Installing Pipe Fittings, Valves, and Flanges
(19107)

Module Six
Installing Fiberglass Pipe Insulation
(19106)

Module Five
Chilled and Hot Water Heating Systems
(19210)

Module Four
Plumbing Systems
(19201)

Module Three
Characteristics of Pipe
(19105)

Module Two
Material Handling, Storage, and Distribution
(19104)

Module One
Orientation to the Trade
(19101)

Core Curriculum: Introductory Craft Skills

This course map shows all of the modules in *Mechanical Insulating Level One*. The suggested training order begins at the bottom and proceeds up. Skill levels increase as you advance on the course map. The local Training Program Sponsor may adjust the training order.

This page is intentionally left blank.

Orientation to the Trade

OVERVIEW

The mechanical insulation industry offers the opportunity to begin employment while learning to become an insulation mechanic. With a combination of classroom work and on-the-job learning, most individuals can become an insulation mechanic in three years.

Module 19101

Trainees with successful module completions may be eligible for credentialing through the NCCER Registry. To learn more, go to **www.nccer.org** or contact us at 1.888.622.3720. Our website has information on the latest product releases and training, as well as online versions of our *Cornerstone* magazine and Pearson's product catalog.

Your feedback is welcome. You may email your comments to **curriculum@nccer.org**, send general comments and inquiries to **info@nccer.org**, or fill in the User Update form at the back of this module.

This information is general in nature and intended for training purposes only. Actual performance of activities described in this manual requires compliance with all applicable operating, service, maintenance, and safety procedures under the direction of qualified personnel. References in this manual to patented or proprietary devices do not constitute a recommendation of their use.

Objectives

When you have completed this module, you will be able to do the following:

1. Describe insulation types and uses, as well as safety procedures and career opportunities in the mechanical insulation industry.
   a. Describe types of insulation.
   b. Describe why and where insulation is used.
   c. Describe the safety practices applicable to insulation.
   d. Describe the career opportunities available to insulation installers.
2. Describe the craft-specific tools used in insulation work.
   a. Describe craft-specific tools used in the trade.
   b. Explain how to properly handle, maintain, and store craft-specific tools.
3. Understand the apprenticeship/training process for insulation installers.
   a. List Department of Labor (DOL) requirements for apprenticeship.
   b. Describe on-the-job learning (OJL) in training and apprenticeship programs.
4. Understand the responsibilities of the employee and employer.
   a. Identify employee responsibilities.
   b. Identify employer responsibilities.

Performance Tasks

This is a knowledge-based module; there are no Performance Tasks.

Trade Terms

Ambient
Architect
Boilers
Breechings
Condensation
General contractor
Heating, ventilating, and air conditioning (HVAC)
Industrial plant
Mechanic

Mechanical contractor
Mechanical insulator
Mechanical insulation
Occupational Safety and Health Administration (OSHA)
On-the-job learning (OJL)
Personnel protection
Plumbing system

Industry Recognized Credentials

If you are training through an NCCER-accredited sponsor, you may be eligible for credentials from NCCER's Registry. The ID number for this module is 19101. Note that this module may have been used in other NCCER curricula and may apply to other level completions. Contact NCCER's Registry at 888.622.3720 or go to **www.nccer.org** for more information.

Contents

Figures and Tables

This page is intentionally left blank.

SECTION ONE

1.0.0 INTRODUCTION TO INSULATION

Objective

Describe insulation types and uses, as well as safety procedures and career opportunities in the mechanical insulation industry.

a. Describe types of insulation.
b. Describe why and where insulation is used.
c. Describe the safety practices applicable to insulation.
d. Describe the career opportunities available to insulation installers.

Trade Terms

Ambient: Relating to the immediate surroundings of an object, system, or person.

Architect: A person whose profession is to design and create plans for buildings, bridges, and facilities.

Boilers: Pressure vessels in which water is turned into steam for heating purposes, for operating equipment, or to generate power.

Breechings: The ducts or pipes connecting the exhaust-gas discharge from a boiler furnace to the stack.

Condensation: Water droplets formed when humid ambient air comes in contact with something cooler.

General contractor: A firm that usually manages all crafts involved in constructing a facility.

Heating, ventilating, and air conditioning (HVAC): A type of system designed to provide thermal comfort and acceptable air quality. The majority of HVAC systems provide both cooling and heating, although some provide only cooling and others provide only air heating.

Industrial plant: A facility that produces electricity or other products used by the general population, such as gasoline and plastics.

Mechanic: An individual with deep knowledge of the applications of a specific mechanical craft.

Mechanical contractor: A firm that installs systems in commercial/industrial facilities. Some mechanical contractors are trade-specific, some do plumbing and HVAC, some plumbing only, and some HVAC only.

Mechanical insulation: A system of materials applied to piping, ductwork, and equipment to protect personnel, and to limit the transmission of heat or noise.

Mechanical insulator: A craftworker who applies and installs mechanical insulation systems.

Personnel protection: Insulation applied only where employees may come in contact with hot pipes or equipment.

Plumbing system: A system of pipes, valves, and fittings that carries water to and drains wastewater from a building, facility, or home.

A mechanical insulation mechanic installs insulation on commercial and industrial projects. Work is completed during the initial construction phase, remodeling or plant modifications, and ongoing maintenance programs. An insulation mechanic performs work on piping, tanks, vessels and equipment for plumbing, utility, heating and air condition systems, and other processing systems.

1.1.0 Types of Insulation

There are several types of mechanical insulation, each of which is manufactured to serve one or more of the following purposes:

- Maintain operating temperatures for industrial processes
- Maintain a comfortable ambient temperature
- Personnel protection
- Maintain a safe ambient noise level

Insulations are categorized into five groups: cellular, fibrous, flake, granular, and reflective. Insulation can be further categorized by its appropriate temperature range. There are also some times of specialty insulation designed for unique applications.

1.1.1 Cellular Insulations

Cellular insulations are comprised of small cells in a structure that may be composed of glass, various kinds of plastics, or rubber. Cellular insulation may be further subdivided as open cell (*Figure 1*), in which the cells interconnect, or closed cell, in which the cells are sealed from one another. While all cellular insulations have some percentage of connected cells, a material with greater than 90 percent closed-cell content is considered closed-cell.

Figure 1 Cellular glass insulation in a variety of shapes.

Cellular insulation is typically manufactured in block and pipe forms. Common types of cellular insulation include the following:

- Elastomeric and polyolefin (often called *flexible insulation*)
- Foam

  - Expanded polystyrene foam (or EPS)
  - Extruded polystyrene foam (or XPS)
  - Polyisocyanurate (or PIR)
  - Phenolic (often called *rigid foam insulation*)
  - Alternative foams:
    ~ Melamine
    ~ Polyimide
- Cellular glass

1.1.2 Fibrous Insulations

Fibrous insulations consist of small-diameter fibers that finely divide the air space within the insulation material. They are typically, but not always, held together by a binder.

Fibrous insulations are further classified as either wool- or textile-based insulations. One form of fibrous insulation, called *glass mineral wool*, is shown in *Figure 2*. Textile-based product is composed of fibers and yarns, which may be woven or non-woven, and organic or inorganic. These materials are sometimes supplied with a coating or as composites created for their specific properties, such as weather resistance, chemical resistance, or reflectivity.

Fibrous insulations are manufactured in pipe, blanket, and board forms. Typical fibrous insulations include the following:

- Fiberglass
- Mineral fiber (or mineral wool)

Figure 2 Glass mineral wool.

- Textile glass
- High temperature fiber

WARNING!

Asbestos is a hazardous substance once commonly used as an insulating material. Asbestos was also regularly used in other applications requiring a resistance to high temperatures, such as automotive brake linings. In 1975, the use of asbestos in construction materials in the United States began to be severely restricted. However, you may encounter abatement programs for the removal of asbestos insulation from older facilities. If you work on such a program, you will be provided the required safety training, equipment, and procedures.

1.1.3 Flake Insulations

Flake insulation is composed of small particles or flakes that finely divide the air space within the insulation material. In some systems, these flakes are bonded together. In other systems, the flakes are loose.

Vermiculite is a flaked insulation and filler material formed by heating a mica-like mineral in a furnace. When heated, the layers greatly expand, trapping air between them. In a coarse form, it can be used for masonry block filler (*Figure 3*) or finely ground and mixed with an adhesive binder, which is sprayed as a fireproof coating for building structures.

1.1.4 Granular Insulations

Granular insulations are composed of small nodules that contain voids or hollow spaces. Granular materials are sometimes considered open-cell materials, since gases can be transferred between the individual spaces. Calcium silicate (*Figure 4*) and molded perlite are examples of granular insulations.

Granular insulations are manufactured in pipe, block, and blanket forms. Common types of granular insulation include the following:

- Calcium silicate
- Perlite
- Microporous
- Aerogel granule composites

1.1.5 Reflective Insulations

Reflective insulations (*Figure 5*) are treatments added to surfaces to reduce heat emittance, which in turn reduces the tendency to transmit heat from outside the facility. Some systems consist of multiple layers that are spaced to minimize convective heat transfer.

Figure 3 Coarse vermiculite.

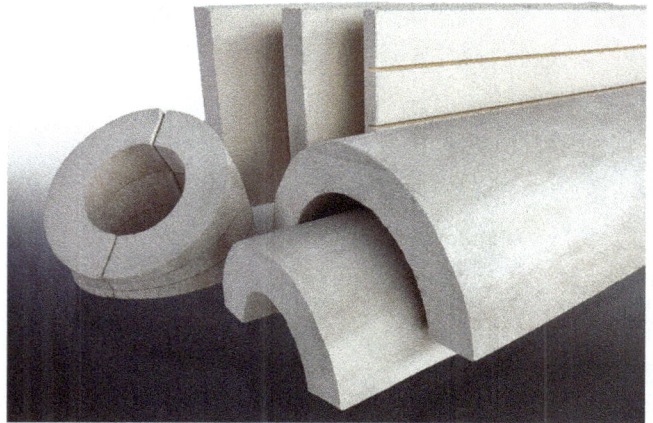

Figure 4 Calcium silicate.

Another class of materials, called *thermal insulating coatings*, falls into this category. Thermal insulating coating is a heat-reflective paint used on pipes, ducts, and tanks.

1.1.6 Categorization by Temperature

Composition and structure are not the only ways to categorized insulation. It can also be categorized by the temperature range of its intended use, as illustrated in *Table 1*.

1.1.7 Specialty Insulation

There are several types of specialty insulation. Two of the most common are the following:

- *Grease duct insulation* – Cooking oils form aerosols during some cooking operations. These oil aerosols are highly flammable. Grease ducts are installed in the ventilation systems near restaurant stovetops to collect the aerosols; the ducts are cleaned on a regular basis that is defined by local laws. The ducts are insulated against fire in case the aerosols are ignited.
- *Refractory* – Some processes require prolonged high temperatures. Equipment associated with those processes require insulation that can withstand those temperatures, which are typically around 1,000°F (538°C). Such equipment includes incinerators, reactors, boilers, and kilns. Materials that can withstand these conditions are called *refractory materials*, and some are used as insulation.

1.1.8 Fabrication of Insulation Systems

Most mechanical insulation systems require some degree of fabrication. The amount of fabrication required will vary depending on the complexity of the job and the materials used. Some mechani-

Figure 5 Reflective insulation.

cal insulation products can be ordered directly from insulation manufacturers in standard sizes with factory-applied facings. These products still require some fabrication in the field to accommodate valves, fittings, and other features. Other insulation materials (for example, cellular glass, phenolic, polyisocyanurate, and polystyrene) are manufactured in relatively large-size blocks or buns, and must be cut into the appropriate size and shape. While this fabrication is sometimes done in the field, the work is more often done by insulation fabricators or distributor/fabricators that specialize in this work.

1.2.0 Uses of Insulation

Insulation is used in a variety of industrial and commercial applications. Facilities install insulation for a variety of reasons, including the following:

- Conserving energy
- Protecting staff from hot surfaces and steam (personnel protection)

Table 1 Categories of Thermal Insulation and Their Temperatures

Temperature Range	Category
Below –50°F (–46°C)	Cryogenic
–50°F to +75°F (–46°C to +24°C)	Refrigeration
+75°F to +1,200°F (+24°C to +649°C)	Medium to high temperature
Above +1,200°F (649°C)	Refractory

Aerogels: Solid Smoke

An aerogel is a man-made material structured like gelatin, except that the liquid has been replaced by air. Microscopically, extremely tiny solid particles join in branching connections spanning voids that are about a hundred times larger in volume. The result is a rigid solid that is extremely light and highly resistant to the transmission of heat.

In powdered form, it can be used as a paint thickener and as a lightweight component of composite materials. Carbon aerogels are being proposed for aircraft wing de-icers because they can carry a small electric current. As insulation, they can be made as planks, broken into small particles and mixed with fibers to make flexible blankets and fabrics, or powdered and combined with a hardening liquid matrix. The applications of aerogels continue to expand into many unusual areas, even catching comet dust during interplanetary space missions.

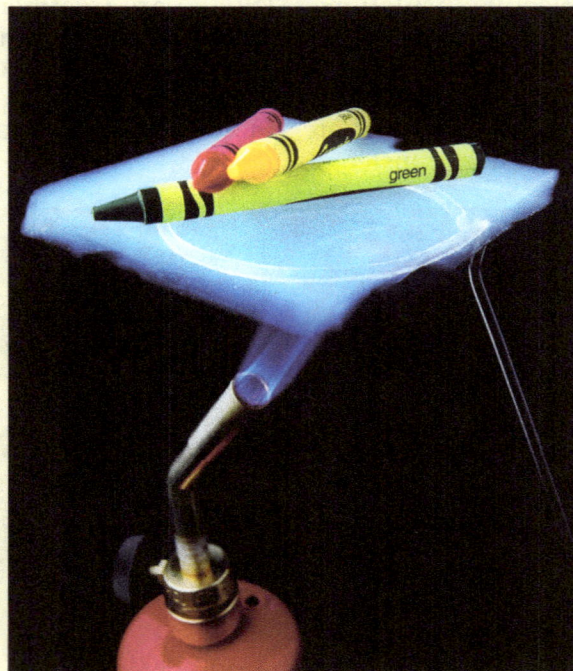

- Maintaining manufacturing processing temperatures
- Reducing fire hazards
- Reducing noise
- Preventing moisture and condensation damage
- Reducing emission of air pollutants, which protects the environment
- Reducing requirements for air conditioning and heat generating systems

1.2.1 Forms of Insulation and Their Uses

Most of the categories of insulation contain several forms of insulations. Each of these has been developed for one or more specific uses. The following is a list of some of the most common forms and their applications:

- *Loose fill* – Used for pouring or blowing in attic or wall spaces of facilities. It is also used to fill minor voids and cracks, and line the walls of chiller boxes.
- *Batts* – A loose, fluffy insulation material with little structural strength that must be placed in an enclosure or laid flat. Fiberglass, the most common form of batt insulation, is also one of the most common types of insulation in general.
- *Blanket* – A flexible material used to wrap different shapes and forms (example: duct wrap).
- *Semi-rigid boards* – Sheets and pre-formed shapes with little flexibility or give (example: fiberglass boards and pipe insulation).
- *Flexible* – Plastic sheets and tubing insulation used in various applications such as refrigerant pipe on residential air conditioning systems (example: Armaflex).

- *Rigid boards* – Block, sheets, and pre-formed shapes with no flexibility, or give, used on straight sections of flat surfaces such as ducts. May be fabricated for curved and irregular surfaces (example: calcium silicate/expanded perlite pipe and masonry block insulation).
- *Tapes* – Used to wrap small diameter apparatus where other forms are not practical (example: Armaflex or Tetraglass).
- *Cements* – Used for molding various shapes and surfaces (example: one shot or finishing).
- *Foam-in-place plastic materials* – A liquid mixed at the time of application. It sets or hardens to insulate surfaces or cavities (example: foamed-in-place urethane).
- *Spray-on fiber, granular, or cement materials* – Used for facility and equipment insulation in a wide temperature application range (example: spray-on fireproofing).
- *Reflective insulation* – Layers of reflective stainless steel foil. It is used extensively in nuclear power plants. Another type is reflective glass or film for plate glass windows that reflect the heat from the sun.

1.2.2 Removable Insulation

Many of the systems to which insulation is applied require periodic maintenance. Some systems fail and require repair. Removable/reusable insulation (*Figure 6*) allows an insulation system to be removed and set aside while maintenance or repair takes place, and then reinstalled with no loss of insulation efficiency and no need for new insulation product.

Removable insulation is generally made of a flexible insulating material that is enclosed with a tough fabric or metal mesh. The seams are usually machine-sewn with heavy-duty thread, staples, or metal rings, commonly called hog rings.

Removable insulation sometimes takes the form of blankets, but is also often a custom product made to fit a specific application.

1.3.0 Insulation Safety

Insulating materials are made of a wide variety of manufactured materials and chemicals. Many types of insulation consist of fine fibers or tiny particles that can be easily breathed in or can enter the skin by punctures or abrasions. Some adhesives and mastics and/or the vapors they give off are toxic. Every person reacts differently to the entry of such materials into the body, so it's important to protect yourself from their unhealthy effects.

(A) FASTENED ATTACHING FEATURES

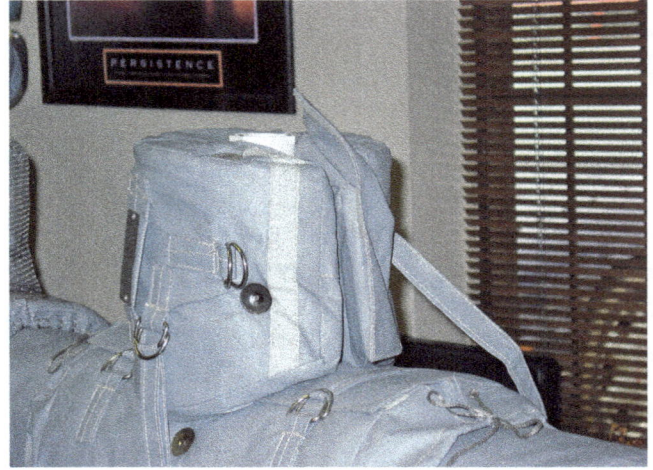

(B) UNFASTENED ATTACHING FEATURES

Figure 6 Removable/reusable insulation.

- Always carefully review the safety data sheet (SDS) for specific safety requirements for the material you'll be working with before the job starts. The National Insulation Association and its members strongly recommend following all safe work practices while handling, working with, and/or installing, removing, or disposing of any insulation products.
- Wear any personal protective equipment (PPE) listed on the SDS for the material you are working with and follow all instructions provided by your employer.
- PPE should fit properly, as prescribed by the PPE manufacturer, the regulations of the job site, and any additional safety guidelines. In particular, respiratory PPE should fit snugly against the face. Facial hair tends to interfere with the proper fit for some respirators.
- It is a federal occupational health requirement to maintain sufficient and proper mechanical or natural ventilation to ensure concentrations of harmful airborne substances are kept at lower levels. Even nuisance dust should be removed by good ventilation.

- Protect your eyes from dust and particles by wearing appropriate eye protection when handling any insulation materials. Such eye protection includes, but is not limited to safety glasses with side shields, goggles, or face shields. Wearing long-sleeved, loose fitting clothes, head covering, and gloves may also be recommended.

1.4.0 Career Paths

Insulation is used in almost every construction project, but not all insulation is mechanical insulation. Mechanical insulation is used on pipes, ducts, and equipment. Structural insulation is used to insulate entire buildings. While it contributes to thermal comfort, it also helps control temperature ranges for industrial and manufacturing processes and helps keep personnel safe.

There are several entry points to become an insulation mechanic. A prospective mechanic might start as a helper or worker, performing the basic work of moving and storing materials and ensuring that the work area is clean. If a helper shows initiative, he or she might go on to become a trainee and begin learning the mechanical insulation trade.

A trainee may then choose to enter into a registered apprenticeship program to gain portable credentials that certify his or her knowledge. These credentials can be used as evidence of the insulation mechanic's abilities to a variety of potential employees.

Apprentices proceed through several levels of training and testing. Generally, apprentices are referred to by the number of years they have completed (*first year apprentice*, *second year apprentice*, etc.). Most mechanical insulation apprenticeships last 3 years, although 4 years is not uncommon. Successful completion of an apprenticeship leads to the level of insulation mechanic or journeyman.

A successful insulation mechanic can go on to one or more of several supervisory positions:

- *Crew leader / foreman* – The crew leader / foreman supervises a group of installers.
- *General foreman, supervisor, superintendent, or site superintendent* – These individuals work with and supervise other skilled trades along with insulation mechanics.
- *Estimating, budgeting, scheduling* – An estimator works with managers and/or owners to prepare quotes and estimates.

Some insulation mechanics cross over to other disciplines. A common crossover is to the related discipline of heating, ventilating, and air conditioning (HVAC). A knowledgeable mechanical insulator may also find work as an instructor in either a high school or a technical college.

1.4.1 Industrial and Commercial

Mechanical insulation can be divided into two broad classifications: industrial and commercial.

Industrial mechanical insulation consists of insulation applied to pipe, duct, and equipment, as well as insulation applied in manufacturing plants, power plants, ships, refineries, chemical plants, and other industrial facilities. A facility may only require mechanical insulation on the HVAC system and plumbing system. Other facilities also require insulation on tanks, process pipes, vessels, boilers, and breechings. Some industrial contractors participate in specialized fields of insulation, especially in low-temperature systems and foam-in-place applications.

In industrial work, the mechanical contractor or general contractor may subcontract work to an insulation contractor or the insulation contractor may contract directly with the plant. Except for new construction or installation, most industrial work is directly contracted with the plants, usually through the plant engineer.

Commercial work is performed in facilities other than an industrial plant. For example, office buildings have plumbing, which consists of cold water, hot water, and drain pipes. They also have HVAC systems consisting of pipes and ducts. Pipes and ductwork must be insulated to hold heat in hot pipes and/or ducts, and to prevent condensation (sweating) on cold pipes or ducts while keeping them cold.

1.4.2 Mechanical Insulation Employment

Mechanical insulation work can be found on both new construction projects and on facilities requiring maintenance and/or repair of mechanical insulation components.

Mechanical insulators might be contractors, either to a mechanical insulation company or to the owners or general contractors of a project. On the other hand, many mechanical insulation firms keep insulation mechanics on staff, moving them from project to project as necessary. Mechanical insulators are occasionally assigned to long-term projects such as ongoing inspection and maintenance of facility insulation.

1.4.3 Working Conditions

Candidates for training programs in insulation should expect the following working conditions:

- Work may be indoors or outdoors.
- Projects may require only one person or a team.
- Work may be in spaces that may be described as the following:
  - Hot or cold and/or subject to sudden temperature changes
  - Confined
  - Noisy
  - Dirty
  - Wet or dry
  - Humid
  - Odorous
  - Dangerous (including exposure to hot surfaces and steam, and close proximity to large moving objects)

GOING GREEN

Mechanical Insulation

Mechanical insulation helps maintain correct temperatures for gases and liquids flowing through pipes and ducts, keeping heated material warm and chilled material cool. This means that the gases and liquids require less thermal conditioning, which leads to shorter operating cycles for the thermal conditioning equipment. This reduces the overall need for energy, resulting in fewer greenhouse gases being released into the atmosphere and a reduced demand for new power plants.

- Work may be physically demanding, including climbing, standing for prolonged periods, reaching, lifting, handling, crouching, walking, feeling, talking, and hearing.

Unless the mechanic is in a large city, most work will be out-of-town. For this reason, insulation mechanics must have dependable transportation. Trainees are expected to be on the job promptly when work is scheduled to begin. While many jobs are scheduled for a 40-hour week, longer work weeks are not uncommon. Employees provide their own boots.

Trainees are expected to do any job assigned, even the less desirable tasks. Such work might consist of carrying material and working in tight places.

1.4.4 Qualities of a Good Mechanical Insulator

A good mechanic performs quality work even where no one will see it. If insulation is improperly applied on pipe surfaces operating at a low temperature, condensation may occur. The condensation may drip on the ceiling, or on the contents in the facility, or may become a safety hazard. If pipe operating at a high temperature is improperly insulated, there is wasted energy due to heat loss, which might also become a safety hazard. Either problem will cost the facility owner extra money, and the employing contractor might lose future jobs with that client.

Some insulation materials are easier to work with than others. However, all materials should be applied properly. It takes considerable skill to perform the many jobs required in insulation work. Skill is displayed by using all materials in the most effective manner, avoiding waste, being precise in measuring and cutting, being neat in the use of mastics and coatings, and keeping the work area orderly and clean.

There is great personal satisfaction gained by craftsmanship and by performing work as well as anyone in the trade. Insulation work is a specialized trade and is best done by qualified mechanics working for qualified contractors. A good mechanic recognizes the importance of the trade, and performs all work in a skillful, efficient manner. Mechanics are professionals and conduct themselves as such.

1.4.5 New Construction Project Workflow

A typical project begins with a decision to build; those responsible for making that decision vary by project type. For example, the future owner will likely make that decision for a commercial building, while a facility like a hospital might be initiated by a government body.

The first step in constructing a new project will be the selection of an architect. The architect will select engineers, and together they will create plans and detailed specifications for every system in the facility, including materials to be used and methods to be used to install them. These specifications will include insulation details, identifying acceptable materials and how they are to be applied and finished.

Once plans are complete and approved, the facility owner or general contractor begins taking bids from general contractors for the construction of the facility. The general contractor in turn takes bids from subcontractors or specialty contractors who will be responsible for specific parts of the facility, such as the plumbing system or the HVAC system. Typically, mechanical insulation is done by a specialty contractor. Mechanical insulators follow the architect's and engineer's plans to apply insulation throughout the structure.

Upon award of a contract, the insulation contractor proceeds with obtaining materials and scheduling the work. The plumbing and HVAC contractors will be working at the same time, and the general contractor will have a superintendent or foreman on the job. The insulation contractor's employees must work with the other trades during the project.

Additional Resources

The Mechanical Insulation Best Practices Guide, Thermal Insulation Association of Canada. Available at **www.tiac.ca/en/resources/best-practices-guide**

Mechanical Insulation Design Guide, National Institute of Building Sciences. **www.wbdg.org/guides-specifications/mechanical-insulation-design-guide**

National Industrial and Commercial Insulation Standards Manual, Midwest Insulation Contractors Association (MICA).

1.0.0 Section Review

1. Which of the following is a category of insulations?
   a. Natural
   b. Granular
   c. Heat-resistant
   d. Tubular

2. Which of the following is a purpose for which a facility would install mechanical insulation?
   a. Personnel protection
   b. Absorbing light
   c. Retaining humidity
   d. Reducing ultraviolet rays

3. A document that tells you about the safety requirements for working with a substance or product is called a(n) _____.
   a. OSHA readout
   b. OSHA material report
   c. safety data sheet
   d. manufacturer's safety report

4. Which statement concerning the mechanical insulation trade is true?
   a. HVAC contractors will complete their work after insulation contractors.
   b. Plumbing contractors will complete their work after insulation contractors.
   c. Plumbing and HVAC contractors will be working at the same time as insulation contractors.
   d. Plumbing and HVAC contractors must complete their areas before insulation mechanics can go to work on them.

2.0.0 TOOLS FOR MECHANICAL INSULATION

Objective

Describe the craft-specific tools used in insulation work.

a. Describe craft-specific tools used in the trade.
b. Explain how to properly handle, maintain, and store craft-specific tools.

A number of tools are required for installing mechanical insulation. With proper care and maintenance, these tools can last for many years. Insulation mechanics must become experts at using them.

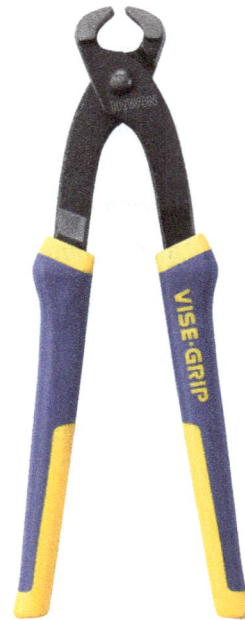

Figure 7 End-cutting nippers.

2.1.0 Tools for the Insulation Mechanic

Without proper tools, a trainee cannot correctly apply insulation or learn the trade. The employer will often provide the special tools necessary for installing insulation.

The following basic tools are required for insulation work; these are covered in more detail in Module 00103 ("Introduction to Hand Tools") from NCCER's *Core Curriculum*:

- Ruler and/or measuring tape
- Utility knife
- Coarse-tooth hand saw
- Keyhole saw

Other tools required for mechanical insulating, including some more trade-specific tools and hand tools, include the following:

- End-cutting nippers (*Figure 7*) (Used to cut wires and pins. Six-inch nippers are good for light gauge wire, eight-inch is a good for general use, and nine-inch nippers are best for heavy gauge wire or large pins.)
- Scissors
- Pointing trowel (*Figure 8*) (used to spread material in tight areas)
- Tool pouch
- Staple gun
- Plastering trowel (*Figure 9*) (used to apply cement and mastic to large, flat surfaces)
- Sheet metal shears (*Figure 10*) (Used to cut metal jacketing by hand. Sheet metal shears are available in three basic styles: M1, M2, and V-19.)

Figure 8 Pointing trowel.

- Paint brush
- Pistol grip bander (*Figure 11*) and/or ratchet bander (used to secure banding material in place for the outer jacketing and/or insulation material)
- Sheet metal awl (*Figure 12*) (Used to scribe lines in metal jacketing to help make straight cuts. Sheet metal awls are also called *scratch awls* or *scribe markers.*)
- Dividers (*Figure 13*) (Used to mark cuts on insulation as well as cut-outs on jacketing. Dividers are useful for marking equal distances, and for laying out semicircles, arcs, and complete circles.)
- Rubber bands

Figure 9 Plastering trowel.

Figure 12 A metal awl.

Figure 13 Sheet metal divider.

- Mallets (*Figure 14*) (used to help fit rigid insulation to large diameter piping and/or equipment by tapping the insulation into place before tightening the securement)
- Hex mesh wire needle (*Figure 15*) (Used to lace poultry netting together prior to applying insulation cement. This tool is also used as and called a *cotter key extractor.*)
- Squeegee for flattening tape
- Lockable tool box

PPE (such as hard hats, safety harnesses, goggles, and shields) is sometimes provided by the employer. Check with your employer to verify what they provide and what you will have to furnish.

Figure 10 Sheet metal shears.

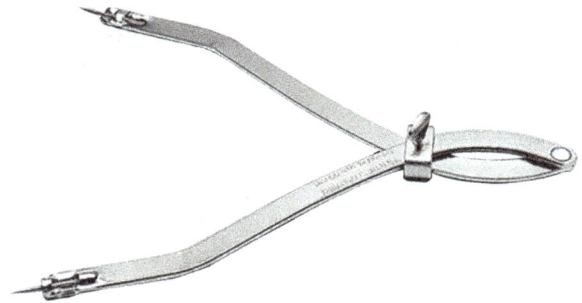

Figure 14 A variety of mallets.

Figure 11 Pistol grip bander.

Figure 15 Hex mesh wire needle.

In addition to the listed hand tools, trainees should become familiar with the following stationary tools used for working with sheet metal:

- Edger (*Figure 16*) (Used to create a $\frac{3}{16}$" flange on the edge of metal jacketing, which is then used as a waterproof seal. This tool is also called an *Easy Edger* and an *E-Z Edger*.)
- Beader, crimper, and former (*Figure 17*) (used to create a bead (or groove) for rigidity and/or to crimp a pattern caps and gores)
- Brake (*Figure 18*) (Used to bend sheet metal. Some brakes are limited to a single fold while others are capable of more complex shapes such as a box. This tool is also called a *bending machine* or *bending brake*.)
- Jacketing shears (*Figure 19*) (used to cut lengths of metal jacket)
- Roller (*Figure 20*) (Used to form sheet metal into curving shapes. The roller can only form a simple curve; complex curves require another tool called an *English Wheel*. A roller is also called a *roll bender*.)

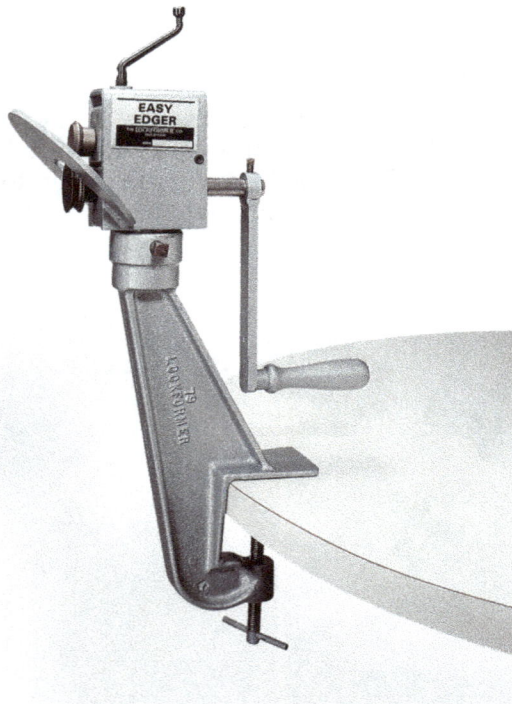

Figure 17 Sheet metal beader/crimper.

Figure 16 Sheet metal edger.

Figure 18 Sheet metal brake.

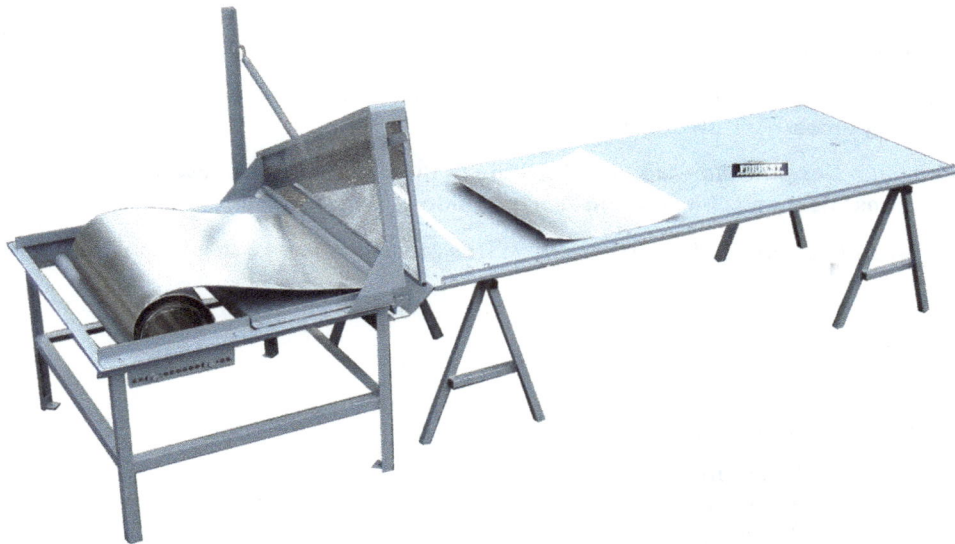

Figure 19 Sheet metal jacketing shears.

2.2.0 Maintenance and Storage

After using insulation tools, always follow these guidelines:

- Keep your tools clean. If necessary, dry the tool after use and before storage.
- Store tools in a dry place. If rust forms on a tool, clean it with a wire brush or steel wool, then coat the tool with a thin film of machine oil.
- Do not abuse your tools. Specifically, do not push them past their design capacity. If you do not have the correct tool for a job, get the correct tool instead of testing the limits of an incorrect tool.
- Use all tools safely, in accordance with the manufacturer's design, federal, state and local safety requirements, and safety regulations for your job site.
- If a tool is damaged, missing components, or not functioning as it should, replace it.

Figure 20 Sheet metal roller.

Additional Resources

The Mechanical Insulation Best Practices Guide, Thermal Insulation Association of Canada. Available at **www.tiac.ca/en/resources/best-practices-guide**

Mechanical Insulation Design Guide, National Institute of Building Sciences. **www.wbdg.org/ guides-specifications/mechanical-insulation-design-guide**

National Industrial and Commercial Insulation Standards Manual, Midwest Insulation Contractors Association (MICA).

2.0.0 Section Review

1. Which of the following is a use for dividers?
   a. Marking a circle to be cut in metal jacket.
   b. Separating insulation by weight.
   c. Splitting insulation by volume.
   d. Creating space in insulation for pipe obstructions.

2. Which of the following statements about caring for tools is true?
   a. Tools that come into contact with oils must be cleaned immediately with soap and water.
   b. Tools that are missing components should only be used with the supervision of your mentor.
   c. Tools that are damaged or have missing components should be replaced.
   d. Tools can be stored without being cleaned as long as they are kept dry.

3.0.0 TRAINING AND APPRENTICESHIP

Objective

Understand the apprenticeship/training process for insulation installers.

    a. List Department of Labor (DOL) requirements for apprenticeship.
    b. Describe on-the-job learning (OJL) in training and apprenticeship programs.

Trade Term

On-the-job learning (OJL): Job-related learning an apprentice acquires while working under the supervision of journey-level workers. Also called *on-the-job training (OJT)*.

Some prospective mechanical insulators begin a training program in high school. Generally, this program will provide little detailed instruction, but it will prepare the future mechanic with knowledge of mechanical insulation work.

Other insulators come to training by starting as workers or helpers for an insulation contractor. Workers who show initiative and a good work ethic can become trainees, mentored by more senior level insulation mechanics as they learn the trade.

Trainees may choose to enter into an apprenticeship. Apprentices agree to work for a company, and to meet the standards of the local apprenticeship committee, while the company agrees to provide training and mentoring to the apprentice. Apprenticeship training is a combination of on-the-job learning (OJL) and classroom instruction.

Apprentice training goes back thousands of years, and its basic principles have not changed. First, it is a means for a person entering the craft to learn from those who have mastered the craft. Second, it focuses on learning by doing—real skills versus theory. Some theory is presented in the classroom. However, it is always presented in a way that helps the trainee understand the purpose behind the skills they are learning.

Registered apprenticeship programs are governed by the Federal Department of Labor and/or state laws and regulations that define in detail how these programs are to be structured and administered. Registered apprenticeship programs do the following:

- Ensure the standards of both mentorship and training
- Protect the safety and welfare of apprentices
- Issue nationally-recognized and portable certificates of completion to apprentices
- Ensure the minimum abilities of program graduates

Apprentices must sign an apprenticeship agreement when entering an apprenticeship program. The employer agrees to provide the training, and the apprentice agrees to meet the program standards established by his or her local apprenticeship committee. The agreement also specifies wage increases as the apprentice meets certain program milestones. Upon completion of the apprenticeship, the trainee will receive certified credentials that can be used to show his or her level of expertise to any potential employer.

3.1.0 DOL Apprenticeship Standards

The US Department of Labor (DOL) Office of Apprenticeship sets the minimum standards for training programs across the country. These programs rely on mandatory classroom instruction and OJL. They require at least 144 hours of classroom instruction per year and 2,000 hours of OJL per year.

Did You Know?

The use of insulation began with the earliest human civilizations: wool and skins for clothing, and wood, stone, and earth for building materials.

Ancient Romans used cork for insulation, including use in shoes to keep their feet warm. Cork was later used to line ice houses, and was an early mechanical insulation material as well.

Man-made insulating fibers date from the industrial period. They were made by injecting steam into molten slag, a waste product from iron furnaces.

NCCER uses the minimum DOL standards as a foundation for comprehensive curricula, providing trainees with in-depth classroom and OJL experience. For mechanical insulation, the Registered Apprenticeship program is 3 years in length. It includes a minimum of 2,000 hours of OJL and 144 hours of related classroom instruction.

This NCCER curriculum provides trainees with industry-driven training and education. It adopts a purely competency-based teaching approach. This means that trainees must show the instructor that they possess the knowledge and skills needed to safely perform the hands-on tasks that are covered in each module.

When an NCCER-certified instructor is satisfied that a trainee has the required knowledge and skills for a given module, that information is sent to NCCER and kept in the Registry system. NCCER's Registry system can then verify training and skills for workers as they move from state to state, from company to company, or even within a company. See the *Appendix* for examples of the credentials issued by NCCER.

Whether you enroll in an NCCER program or another apprenticeship program, make sure you work for an employer or sponsor who supports a nationally standardized training program that includes credentials to confirm your skill development.

3.2.0 On-the-Job Learning

All apprenticeship standards prescribe certain work-related or on-the-job learning (OJL). This training may begin after graduation from high school or before graduation as a part of a youth apprenticeship program.

The OJL is broken down into specific tasks in which the apprentice receives hands-on training. In addition, a specified number of hours is required in each task. The total number of OJL hours for an apprenticeship program is traditionally 8,000, which amounts to four years of training.

In a competency-based program, it may be possible to shorten this time by testing out of specific tasks through a series of performance exams. In a traditional program, the required OJL may be acquired in increments of 2,000 hours per year.

The apprentice must log all work time and turn it in to the apprenticeship committee so that accurate time control can be maintained. After each 1,000 hours of related work, the apprentice typically will receive a pay increase as prescribed by the apprenticeship standards.

For those entering an apprenticeship program, a high school or technical school education is desirable. Courses in shop, mechanical drawing, and general mathematics are helpful. Manual dexterity, good physical conditioning, and quick reflexes are important. The ability to solve problems quickly and accurately and to work closely with others is essential. You must also maintain a high awareness of safety concerns.

A prospective apprentice may be required to submit certain information to the apprenticeship committee such as the following:

- Aptitude test (General Aptitude Test Battery or GATB Form Test) results (usually administered by the local Employment Security Commission)
- Proof of educational background (candidate should have school transcripts sent to the committee)
- Letters of reference from past employers and friends
- Proof of age
- If the candidate is a veteran, a copy of Form DD214
- A record of technical training received that relates to the construction industry and/or a record of any pre-apprenticeship training

NOTE

Some companies have physical activity requirements that apprentices must meet. These requirements vary from company to company. Generally, apprentices must do the following:

- Wear proper safety and personal protective equipment (PPE) on the job
- Purchase and maintain tools of the trade as needed and required by the contractor
- Submit a monthly on-the-job training report to the committee
- Report to the committee if a change in employment status occurs
- Attend classroom-related instruction and adhere to all classroom regulations, such as attendance requirements

Note that informal OJL provided by employers is usually less thorough than OJL provided through a formal apprenticeship program. The degree of training and supervision in this type of program often depends on the size of the employer. A small contractor may provide training in only one area, whereas a large company may be able to provide training in several areas.

Additional Resources

29 *CFR* 1900–1910, **www.ecfr.gov**

29 *CFR* 1926, **www.ecrf.gov**

3.0.0 Section Review

1. The NCCER mechanical insulating training program applies DOL standards and is a _____.

    a. one-year program
    b. two-year program
    c. three-year program
    d. four-year program

2. A prospective apprentice may be required to submit certain information to the apprenticeship committee, such as _____.

    a. Proof of educational background
    b. license
    c. 100 hours of OJL
    d. OSHA 40-hour course

4.0.0 EMPLOYEE AND EMPLOYER RESPONSIBILITIES

Objective

Understand the responsibilities of the employee and employer.
a. Identify employee responsibilities.
b. Identify employer responsibilities.

Trade Term

Occupational Safety and Health Administration (OSHA): The federal government agency established to ensure a safe and healthy environment in the workplace.

The safe and cost-effective installation and maintenance of mechanical insulation requires close collaboration between the employer and the employees. All workers must understand the responsibilities of providing a safe and productive workplace.

4.1.0 Employee Responsibilities

To be successful, you must be able to use current trade materials, tools, and equipment to finish the task quickly and efficiently. You must keep up-to-date on technical advancements and continually gain the skills to use them. A professional never takes chances with regard to personal safety or the safety of others.

4.1.1 Professionalism

The term *professionalism* broadly describes the desired overall behavior and attitude expected in the workplace. Professionalism is too often absent from the construction site and various trades. Many would argue that professionalism must start at the top in order to be successful. It is true that management support of professionalism is important to its success in the workplace, but it is just as important that individuals recognize personal responsibilities for professionalism.

Professionalism includes honesty, productivity, safety, civility, cooperation, teamwork, clear and concise communication, being on time, and coming prepared to work. It can be demonstrated in a variety of ways every minute you are on the job.

Professionalism is a benefit to both the employer and the employee. It is a personal responsibility. The construction industry is what each individual chooses to make of it—choose professionalism and the industry image will follow.

4.1.2 Honesty

Honesty and personal integrity are important traits of successful professionals. Professionals pride themselves on performing a job well and on being punctual and dependable. Each job is completed in a professional way, never by cutting corners or reducing materials. A valued professional maintains work attitudes and ethics that protect tools, materials, and other property belonging to employers, customers, and other trades from damage or theft at the shop or job site.

Honesty and success go hand-in-hand for both the employer and the professional mechanical insulator. It is not simply a choice between good and bad but a choice between success and failure. Dishonesty will always catch up with you. Whether you steal materials, tools, or equipment from the job site or simply lie about your work, it will not take long for your employer to find out.

If you plan to be successful and enjoy continuous employment, consistency of earnings, and being sought after as opposed to seeking employment, then start out with the basic understanding of honesty in the workplace. You will reap the benefits.

Honesty means more, however, than simply not taking things that do not belong to you. It also means giving a fair day's work for a fair day's pay. Employers place a high value on employees who display honesty.

4.1.3 Loyalty

Employees expect employers to look out for their interests, to provide them with steady employment, and to promote them to better jobs as openings occur. Employers feel that they, too, have a right to expect loyalty from their employees— to keep their interests in mind, to speak well of them to others, to keep any minor troubles strictly within the plant or office, and to keep absolutely confidential all matters that pertain to the business. Both employers and employees should keep in mind that loyalty is not something to be demanded; rather, it is something to be earned.

4.1.4 Willingness to Learn

Every company and job site has its own way of doing things. Employers expect their workers to be willing to learn these ways. You must be willing to adapt to change and learn new methods and procedures as quickly as possible. Sometimes, a change in safety regulations or the purchase of new equipment makes it necessary for even experienced employees to learn new methods and operations. Successful people take every opportunity to learn more about their trade.

4.1.5 Taking Responsibility

Most employers expect their employees to see what needs to be done and do it. After an assignment is received and the procedure and safety guidelines are fully understood, you should assume the responsibility for that task without further reminders.

4.1.6 Cooperation

To cooperate means to work together. In our modern business world, cooperation is the key to getting things done. Learn to work as a member of a team with your employer, supervisor, and fellow workers in a common effort to get the work done efficiently, safely, and on time.

4.1.7 Rules and Regulations

Employees can work well together only if there is some understanding about the nature of work to be done, when and how it will be done, and who will do it. Rules and regulations are a necessity in any work situation and must be followed by all employees.

4.1.8 Tardiness and Absenteeism

Tardiness means being late for work, and absenteeism means being off the job for one reason or another. While occasional absences are unavoidable, consistent tardiness and frequent absences are an indication of poor work habits, unprofessional conduct, and a lack of commitment.

Although workers may not be paid when they are absent or tardy, there is still a cost to the employer. For example, the worker's health care insurance must still be paid, even though the worker is not on site. In addition, jobs are bid and scheduled are based on a certain workforce size. If you are not there, work is not being done and schedules are not being met. It is important for you to be at work, on time, every day. If you must be absent, call in as soon as possible so that your employer can find a replacement.

4.1.9 Safety

In exchange for the benefits of your employment and your own well-being, you are obligated to work safely. You are also obligated to make sure anyone you supervise or work with is working safely. Your employer is obligated to maintain a safe workplace for all employees. Safety is everyone's responsibility.

You have a responsibility to maintain a safe working environment. This means two things:

- Follow your company's rules for proper working procedures and practices.

Ethical Principles for Members of the Construction Trades

Honesty—Be honest and truthful in all dealings. Conduct business according to the highest professional standards. Faithfully fulfill all contracts and commitments. Do not deliberately mislead or deceive others.

Integrity—Demonstrate personal integrity and the courage of your convictions by doing what is right even if there is pressure to do otherwise. Do not sacrifice your principles because it seems easier.

Loyalty—Be worthy of trust. Demonstrate fidelity and loyalty to companies, employers and sponsors, co-workers, trade institutions, and other organizations.

Fairness—Be fair and just in all dealings. Do not take undue advantage of another's mistakes or difficulties. Fair people are open-minded and committed to justice, equal treatment of individuals, and tolerance for and acceptance of diversity.

Respect for others—Be courteous and treat all people with equal respect and dignity.

Obedience—Abide by laws, rules, and regulations relating to all personal and business activities.

Commitment to excellence—Pursue excellence in performing your duties, be well-informed and prepared, and constantly try to increase your proficiency by gaining new skills and knowledge.

Leadership—By your own conduct, seek to be a positive role model for others.

- Report any unsafe equipment and conditions directly to your supervisor.

If you see something unsafe while on the job, report it! Do not ignore it. It will not correct itself. In the end, even if you do not think an unsafe condition affects you, it does. Always report unsafe conditions. Do not think your employer will be angry because your productivity suffers while the condition is being reported. On the contrary, your employer will be more likely to criticize you for not reporting a problem.

> **WARNING!**
>
> For the safety of yourself and others, always report unsafe conditions. Ignoring them could cause danger or harm to workers.

Your employer knows that the short time lost in making conditions safe again is nothing compared with shutting down the whole job because of a major disaster. If that happens, you are out of work anyway. In fact, Occupational Safety and Health Administration (OSHA) regulations require you to report hazardous conditions. This applies to every part of the construction industry. Whether you work for a large contractor or a small contractor, you are obligated to report unsafe conditions.

4.2.0 Employer Responsibilities

Just as the employee has responsibilities on the job, the employer also has responsibilities. These are set out in the *Occupational Safety and Health Act of 1970*. The job of OSHA is to set occupational safety and health standards for all places of employment, enforce these standards, ensure that employers provide and maintain a safe workplace for all employees, and provide research and educational programs to support safe working practices.

OSHA was adopted with the stated purpose to assure as best as possible every worker in the nation with safe and healthful working conditions and to preserve our human resources.

OSHA requires each employer to provide a safe and hazard-free working environment. OSHA also requires that employees comply with OSHA rules and regulations that relate to workers' conduct on the job. To gain compliance, OSHA can perform spot inspections of job sites, impose fines for violations, and even stop work from proceeding until the job site is safe.

According to OSHA standards, you are entitled to on-the-job safety training. Your employer must do the following:

- Show you how to do each job safely
- Provide you with the required personal protective equipment
- Warn you about specific hazards
- Supervise you for safety while performing the work

The enforcement for this act of Congress is provided by the federal and state safety inspectors, who have the legal authority to impose fines for safety violations. The law allows states to have their own safety regulations and agencies to enforce them, but the US Secretary of Labor must first approve the states' programs. In states that do not develop such regulations and agencies, federal OSHA standards are mandatory.

OSHA standards are listed in 29 *CFR* 1926, *OSHA Safety and Health Standards for the Construction Industry* (sometimes called *OSHA Standard 1926*). Other safety standards that apply to construction are published in 29 *CFR* 1900–1910.

The most important general requirements that OSHA places on employers in the construction industry are as follows:

- The employer must post, in an easily seen area, signs informing employees of their rights and responsibilities.
- The employer must ensure there are no serious hazards on the job site and make sure the workplace complies with OSHA rules and regulations.

- Warning signs, posters, and labels must be posted in all required areas.
- The employer must perform frequent and regular job site inspections of equipment.
- The employer must instruct all employees to recognize and avoid unsafe conditions and to know the regulations that pertain to the job so employees may control or eliminate any hazards.
- No one may use any tools, equipment, machines, or materials that do not comply with 29 *CFR*, Part 1926.
- The employer must ensure that only qualified individuals operate tools, equipment, and machines.
- The employer must provide medical training and examinations when required by OSHA.

- Employers with more than than 10 employees must keep records of work-related injuries and illnesses. These records must be available to employees.
- Employers must not discriminate against employees who are exercising their rights under OSHA regulations.

Additional employer responsibilities are described in the Americans with Disabilities Act of 1990 (ADA). If a worker has a disability but is qualified to do the job, that worker has equal rights to employment. The US Equal Employment Opportunity Commission, along with state and local civil rights agencies, enforce ADA regulations.

Additional Resources

29 *CFR* 1900–1910, **www.ecfr.gov**

29 *CFR* 1926, **www.ecrf.gov**

4.0.0 Section Review

1. Which of the following terms broadly describes the desired overall behavior and attitude expected in the workplace?

   a. Professionalism
   b. Collaboration
   c. Honesty
   d. On-the-job behavior

2. OSHA standards for general industry are covered in _____.

   a. 29 *CFR*, Parts 1900–1910
   b. 29 *CFR*, Part 1970
   c. 29 *CFR*, Part 1926
   d. 29 *CFR*, Parts 1965–1983

SUMMARY

Mechanical insulation is a critical component of industrial and commercial facilities. Insulation maintains thermal conditioning, supports industrial processes, and protects personnel and equipment. A career in mechanical insulation can be satisfying for the person who enjoys confronting problems and devising solutions.

Before you begin work you will need to obtain basic mechanical insulation tools. As you progress, you will also need to learn to use other tools provided by your employer. All tools must be used, cared for, and stored properly for long tool life.

Many jobs in the facility construction and maintenance field are potentially dangerous, including mechanical insulation. Learn and follow all safety instructions pertaining to both mechanical insulation work and to the safe handling of materials used for mechanical insulation. In particular, mechanical insulators must pay attention to rules for respiratory safety when working with fiber-based insulations.

There are many career paths available to insulation mechanics. They can work in the industrial or commercial sectors, or both. There are temporary positions, such as a position supporting new construction, and longer-term positions, such as ongoing maintenance and repair in large operations. One might start out as a worker before becoming a trainee, progressing through several apprenticeship levels to finally become an insulation mechanic or journey-level worker. Those with particular aptitude and drive might go on to help put together project bids and may advance to a supervisory or management position.

1. Which of the following is a reason why pipes and equipment are insulated?
   a. Insulation protects the pipe from exposure to the sun's rays.
   b. Insulating loud machinery helps lower the ambient noise level.
   c. Insulation absorbs ambient humidity making the facility more comfortable.
   d. Insulation protects the equipment from impact damage.

2. Which of the following is no longer used as a major component in insulation?
   a. Fiberglass
   b. Asbestos
   c. Mineral wool
   d. Urethane

3. Insulation used in applications that operate above 1,200°F (649°C) is categorized as _____.
   a. refractory
   b. reflective
   c. high-temperature
   d. ablative

4. The main purpose of grease duct insulation is to _____.
   a. reduce the amount of noise from such systems in restaurants
   b. keep the grease aerosols warm so they don't solidify in the duct
   c. collect the grease aerosols created by cooking
   d. block the heat generated by a grease fire in the duct

5. Which of the following statements is true of insulating tapes?
   a. Tapes are not insulation; they are used to hold two or more pieces of insulation together.
   b. Insulating tapes must be periodically removed and re-applied.
   c. Tapes are a good solution for excessively humid applications.
   d. Tapes are a good solution for small diameter components.

6. A safety data sheet (SDS) dictates the _____.
   a. order of installation
   b. safety requirements for a specific material
   c. allowable amount of waste
   d. amount of material necessary to complete a project

7. Which of the following statements about personal protective equipment (PPE) is true?
   a. It is always your responsibility to provide your PPE.
   b. Some tasks are too simple or quick to necessitate wearing PPE.
   c. Some insulation materials do not have an SDS.
   d. Facial hair can interfere with proper fit of respiratory PPE.

8. What are the two broad categories of mechanical insulation?
   a. Industrial and commercial
   b. Industrial and residential
   c. Commercial and residential
   d. Off-the-shelf and fabricated

9. Which of the following statements is true for a candidate in an insulation training program?
   a. If you are hired as a trainee then you are required to become an apprentice.
   b. Plumbing and HVAC contractors directly supervise mechanical insulators.
   c. You might have to work in hot or cold conditions.
   d. You are required to purchase any tool you may need.

10. Which of the following statements is true?
    a. Hiring an architect is one of the first steps in building a new facility.
    b. A facility owner hires specialty contractors.
    c. A foreman hires a general contractor.
    d. A specialty contractor cannot hire other specialty contractors.

11. Plumbing, electrical, and insulation contractors are known as _____.
   a. architects
   b. general contractors
   c. specialty contractors
   d. utility contractors

12. Which of the following would be in charge of hiring a subcontractor to install a project's plumbing systems?
   a. General contractor
   b. Plumbing contractor
   c. Project superintendent
   d. Architect

13. Which of the following is used to scribe lines in sheet metal?
   a. Utility knife
   b. Snap line
   c. Chalk
   d. Awl

14. If a tool becomes rusty, you should _____.
   a. clean it with a wire brush or steel wool
   b. replace it
   c. return it to the manufacturer
   d. remove the rust with machine oil

15. Minimum standards for apprentice training programs are established by _____.
   a. OSHA
   b. the DOL
   c. NCCER
   d. the employer

16. Which of the following statements about mechanical insulation apprenticeships is true?
   a. Time on the job can be substituted for classroom training.
   b. Every registered apprenticeship program for mechanical insulation is 4 years in length.
   c. Mandatory classroom training is a part of every apprenticeship.
   d. Apprentices may be called on to mentor other apprentices.

17. Which of the following is true with regard to tardiness and absenteeism?
   a. It is never acceptable to be absent, even when you are contagious.
   b. It is okay to be a little late for work as long as you make up the time.
   c. If you must be absent, call in early so that your employer can find a replacement.
   d. If you are not being paid, your absence does not cost the company anything.

18. If you see a safety violation at your job site, you should _____.
   a. ignore it unless it affects you directly
   b. make a mental note to avoid the area in future
   c. report it to your supervisor
   d. assume it is okay as long as no one has been injured

19. The primary mission of OSHA is to _____.
   a. inspect job sites for safety violations
   b. fine companies that violate safety regulations
   c. distribute safety equipment to workers
   d. ensure that employers maintain a safe workplace

20. In a traditional 4-year apprenticeship program how many total hours of on-the-job learning are required?
   a. 2,000
   b. 4,000
   c. 8,000
   d. 10,000

Trade Terms Quiz

Fill in the blank with the correct term that you learned from your study of this module.

1. _____ connect a furnace to a stack.
2. HVAC or plumbing systems are installed by a(n) _____.
3. The _____ transports water to and from a facility.
4. _____ is applied to a facility's mechanical components to preserve a thermal condition or protect personnel.
5. Prevailing conditions surrounding something are referred to as its _____ conditions.
6. A trade professional with deep knowledge of a specific craft is a(n) _____.
7. A craft professional who installs insulation is a(n) _____.
8. Insulation which protects workers from hot surfaces is performing a(n) _____ function.

9. A(n) _____ produces power or products.
10. _____ are pressure vessels in which water is heated.
11. A(n) _____ creates plans for a facility and is usually in charge of the construction.
12. A(n) _____ oversees the work of specialty contractors.
13. _____ forms when water vapor comes into contact with a cool surface.
14. The _____ system carries thermally conditioned fresh air to various parts of a facility.
15. _____ is the federal government agency established to ensure a safe and healthy environment in the workplace.
16. Job-related learning acquired while working is known as _____.

Trade Terms

Ambient
Architect
Boilers
Breechings
Condensation
General contractor

Heating, ventilating, and air conditioning (HVAC)
Industrial plant
Mechanic
Mechanical contractor
Mechanical insulator

Mechanical insulation
Occupational Safety and Health Administration (OSHA)
On-the-job learning (OJL)
Personnel protection
Plumbing system

Trade Terms Introduced in This Module

Ambient: Relating to the immediate surroundings of an object, system, or person.

Architect: A person whose profession is to design and create plans for buildings, bridges, and facilities.

Boilers: Pressure vessels in which water is turned into steam for heating purposes, for operating equipment, or to generate power.

Breechings: The ducts or pipes connecting the exhaust-gas discharge from a boiler furnace to the stack.

Condensation: Water formed when ambient air comes in contact with something cooler.

General contractor: A firm that usually manages all crafts involved in constructing a facility.

Heating, ventilating, and air conditioning (HVAC): A type of system designed to provide thermal comfort and acceptable air quality. The majority of HVAC systems provide both cooling and heating, although some provide only cooling and others provide only air heating.

Industrial plant: A facility that produces electricity or other products used by the general population, such as gasoline and plastics.

Mechanic: An individual with deep knowledge of the applications of a specific mechanical craft.

Mechanical contractor: A firm that installs systems in commercial/industrial facilities. Some mechanical contractors are trade-specific, some do plumbing and HVAC, some plumbing only, and some HVAC only.

Mechanical insulation: The thermal, acoustical, and personnel safety systems applied to piping, ductwork, and equipment.

Mechanical insulator: A craftworker who applies and installs mechanical insulation systems.

On-the-job learning (OJL): Job-related learning an apprentice acquires while working under the supervision of journey-level workers. Also called *on-the-job training (OJT)*.

Occupational Safety and Health Administration (OSHA): The federal government agency established to ensure a safe and healthy environment in the workplace.

Personnel protection: Insulation applied only where employees may come in contact with hot pipes or equipment.

Plumbing system: A system of pipes, valves, and fittings that carries water to and drains wastewater from a building, facility, or home.

SAMPLES OF NCCER TRAINING CREDENTIALS

NCCER

Board of Trustees confers upon

Sample Student

this certificate of completion for

Mechanical Insulating Level One

in the Standardized Craft Training program
on this Fifth day of October, 2017

Donald E. Whyte
Donald E. Whyte
President, NCCER

Student 7 Sample
Certified Plus
4671784

Sample Student
2781481

Additional Resources

This module presents thorough resources for task training. The following reference material is recommended for further study.

The Mechanical Insulation Best Practices Guide, Thermal Insulation Association of Canada. Available at
www.tiac.ca/en/resources/best-practices-guide

Mechanical Insulation Design Guide, National Institute of Building Sciences.
www.wbdg.org/guides-specifications/mechanical-insulation-design-guide

National Industrial and Commercial Insulation Standards Manual, Midwest Insulation Contractors Association (MICA).

29 *CFR* 1900–1910, **www.ecfr.gov**

29 *CFR* 1926, **www.ecrf.gov**

Figure Credits

Courtesy of CertainTeed Insulation, Module opener

Pittsburgh Corning Corporation, Figure 1

Knauf Insulation, Figure 2

©iStockphoto.com/praisaeng, Figure 3

Industrial Insulation Group, Figure 4

Courtesy NASA/JPL-Caltech, SA01

Innovative Energy, Inc., Figure 5

Coverflex Manufacturing, Inc. (US), Figure 6

Courtesy of Irwin Tools, Figures 7, 10

Kraft Tool Co., Figures 8, 9

Midwest Industrial Packaging, Figure 11, Exam Figure 1

Bart van der Maar, Figures 12, 17, 18, 20

Malco Products, Inc., Figure 13

TEKTON®, Figure 14

General Tools, Figure 15

Eagle Morlin Manufacturing, Inc., Figure 16

Forrest Mfg. Co., Figure 19

Section Review Answer Key

Answer	Section Reference	Objective
Section One		
1. b	1.1.0; 1.1.4	1a
2. a	1.2.0	1b
3. c	1.3.0	1c
4. c	1.4.5	1d
Section Two		
1. a	2.1.0	2a
2. c	2.2.0	2b
Section Three		
1. c	3.1.0	3a
2. a	3.2.0	3b
Section Four		
1. a	4.1.1	4a
2. a	4.2.0	4b

This page is intentionally left blank.

NCCER CURRICULA — USER UPDATE

NCCER makes every effort to keep its textbooks up-to-date and free of technical errors. We appreciate your help in this process. If you find an error, a typographical mistake, or an inaccuracy in NCCER's curricula, please fill out this form (or a photocopy), or complete the online form at **www.nccer.org/olf**. Be sure to include the exact module ID number, page number, a detailed description, and your recommended correction. Your input will be brought to the attention of the Authoring Team. Thank you for your assistance.

Instructors – If you have an idea for improving this textbook, or have found that additional materials were necessary to teach this module effectively, please let us know so that we may present your suggestions to the Authoring Team.

NCCER Product Development and Revision
13614 Progress Blvd., Alachua, FL 32615

Email: curriculum@nccer.org
Online: www.nccer.org/olf

❏ Trainee Guide   ❏ Lesson Plans   ❏ Exam   ❏ PowerPoints   Other _____

Craft / Level: _____ Copyright Date: _____

Module ID Number / Title: _____

Section Number(s): _____

Description:

_____

_____

_____

_____

Recommended Correction:

_____

_____

_____

_____

Your Name: _____

Address: _____

_____

Email: _____ Phone: _____

This page is intentionally left blank.

Material Handling, Storage, and Distribution

OVERVIEW

Unloading, storing, and distributing insulation material is a task likely to fall to workers who provide support for insulation mechanics. Insulation can be damaged if not properly handled and stored. Everyone who works with insulation materials must understand proper handling and storage methods.

Module 19104

Trainees with successful module completions may be eligible for credentialing through the NCCER Registry. To learn more, go to **www.nccer.org** or contact us at 1.888.622.3720. Our website has information on the latest product releases and training, as well as online versions of our *Cornerstone* magazine and Pearson's product catalog.

Your feedback is welcome. You may email your comments to **curriculum@nccer.org**, send general comments and inquiries to **info@nccer.org**, or fill in the User Update form at the back of this module.

This information is general in nature and intended for training purposes only. Actual performance of activities described in this manual requires compliance with all applicable operating, service, maintenance, and safety procedures under the direction of qualified personnel. References in this manual to patented or proprietary devices do not constitute a recommendation of their use.

19104 V2

MATERIAL HANDLING, STORAGE, AND DISTRIBUTION

Objectives

When you have completed this module, you will be able to do the following:

1. Describe how to receive and store insulating material.
   a. Describe the receiving procedure.
   b. Describe the proper storage methods for insulating material.
2. Explain how to move and distribute insulating material.
   a. Describe the methods used in moving insulation material.
   b. Describe how to distribute insulating material.

Performance Tasks

This is a knowledge-based module; there are no Performance Tasks.

Trade Terms

All-service jacketing (ASJ)
Butt strips
Construction elevator
Crew
HMIS label
Material storage area
NFPA fire diamond
Packing slip

Pallets
Polyethylene sheeting
Rope and pulley
Scaffolds
Stack
Staging
Tarpaulin

Industry Recognized Credentials

If you are training through an NCCER-accredited sponsor, you may be eligible for credentials from NCCER's Registry. The ID number for this module is 19104. Note that this module may have been used in other NCCER curricula and may apply to other level completions. Contact NCCER's Registry at 888.622.3720 or go to **www.nccer.org** for more information.

Contents

Figures

1.0.0 RECEIVING AND STORING INSULATION MATERIAL

Objective

Describe how to receive and store insulating material.

a. Describe the receiving procedure.
b. Describe the proper storage methods for insulating materials.

Trade Terms

Crew: Mechanics and helpers assigned to a project. Each craft will have a crew for their specific applications.

Material storage area: One or more spaces located at the job site where all materials are stored.

Packing slip: A detailed listing of all materials shipped from a supplier. Every delivery/shipment has a packing slip.

Pallets: Small platforms used to place boxes on, to keep materials off the ground, and to make moving and delivering bulk materials easier.

Polyethylene sheeting: For storage purposes, heavy flexible sheets of plastic used as tarpaulins for protecting materials from the elements.

Scaffolds: Elevated work platforms for both personnel and materials.

Stack: An orderly pile of like materials.

Tarpaulin: Waterproofed fabric or canvas material that is used to cover insulation materials.

P roper receiving and storage methods are crucial to a project. They ensure that the correct materials are received in the amounts ordered for the project. They also ensure that the materials are undamaged between the delivery truck, movement into storage, and movement to the work area.

1.1.0 Delivery and Receiving

On some projects, materials are delivered to the job on a daily or weekly basis as needed. This minimizes or eliminates the need for storing insulating material.

On other projects, all materials are delivered at the beginning of the job. They are stored at the job site until the insulation mechanic installs them. In these situations the materials must be stored in a way that protects them from being damaged or misplaced.

Materials may be delivered from the distributor's warehouse or drop-shipped directly from the manufacturer. Depending on the scale of the project, the materials can be delivered by straight truck or tractor-trailer. Smaller jobs may only require one person to receive and later distribute materials, and larger jobs may have a distribution crew headed by a distribution supervisor.

Every delivery must have a packing slip (also called a *packing list, delivery ticket,* or *bill of lading*), such as that shown in *Figure 1*. The packing slip identifies the contents of the shipment. Sometimes it will also list material which has been backordered.

The packing slip should be carefully checked against material received to verify accuracy. Verify the packing list by comparing it to a copy of the purchase order before signing for the shipment. If there isn't a copy of the purchase order available, the person responsible for the order must be notified, and a note must be made on the shipper's document that delivered materials could not be receipted.

> **WARNING!**
>
> With the exception of hardware, all insulation materials (including insulation, mastics, and adhesives) in a shipment must have an associated safety data sheet (SDS) from the manufacturer. These sheets specify the health and safety issues associated with every item and tell you how to safely handle, store, and install the material. If an SDS is missing, the distributor must be contacted to have it sent to the job site.

1.2.0 Choosing a Storage Area

Ideally, materials should be taken directly from the delivery truck to the area where they are to be used. However, this is often not possible, and most materials must be stored in a material storage area on the job site.

The site superintendent may have a particular area set aside for insulation material storage, or that individual may ask for your feedback when identifying a storage area. Either way, be sure to get the superintendent's approval. Otherwise, you may have to move materials again.

PACKING LIST

Shipped From:
Topaz Insulation Distributor
123 Capella Drive
St Louis, MO 63101

Shipped From:
A1 Insulation Contracting
c/o Smith Industrial Park – Smith Construction Project
917-G Holberg Road
Flint, MO 62223

Gross Weight: 1205 lbs

Shipment Volume: 585 cu ft

Deliver on 05/28/2015	Order # 498132899		Terms Net 30	Order Date 05/20/2015
Item #	**SKU #**	**Product Description**	**Qty Ordered**	**Qty Shipped**
1701	F355-14896	1 × 48 × 96 FL BD SA	60	60
1601	F155-35669	2 × 24 × 48 FL BD SA	40	40
1601-FB	F155-356-FB	2 × 24 × 48 FL BD	20	20

Figure 1 Example packing slip.

The first requirement for the material storage area is size. Insulation materials are bulky and often require considerable storage space. Determine how much material you are likely to have on site at one time. Allow for adequate room to move around in the storage area, and to move materials in and out of the storage area.

Consider the relative locations of the storage area(s) and the work area(s). The two should be as close together as practically possible. Time spent handling and moving materials can have a significant effect on project schedules and profitability.

Finally, the storage area must be dry and not subject to extreme heat or cold. Many insulations and related materials are sensitive to moisture, and adhesives and mastics are sensitive to temperature extremes.

> **NOTE**
> If you are in charge of, or have input to, preparing a bid, take the storage areas and their proximity to the work location into account.

1.2.1 Stacking Guidelines

In storage, organize materials for easy selection to help prevent picking and distributing the wrong material. Place like materials together, and place like sizes together within a given stack. Ensure the product labels face outward and are visible.

When placing material in storage, think of how and when it will be used. Store materials that will be used more frequently and sooner nearer to the entry for easier access; materials that are used less frequently or those scheduled for later should be toward the back of the storage area.

If there are opened boxes, keep them at the front of the stack. Make sure that opened boxes are used before opening a new box (*Figure 2*).

Most insulation materials are susceptible to damage from moisture. Rain might still get into the storage area through incomplete roofing systems and through window openings where windows are not yet installed. Many flooring surfaces are prone to condensation if the humidity is high, and insulation materials and/or their packaging might wick condensation moisture from the floor. All materials should be stored on pallets to protect them from moisture damage.

Keep the following rules of thumb in mind when stacking materials:

- Large stacks of material are more difficult to keep dry. A fabric or canvas tarpaulin (or *tarp*) must cover all of the material and its packaging, and then be secured to the storage pallet. If the tarp cannot cover the entire stack, then separate the stack into smaller stacks. Heavy polyethylene sheeting (e.g., 10 mil thick) may also be used as a tarp. A securely covered pallet is shown in *Figure 3*.

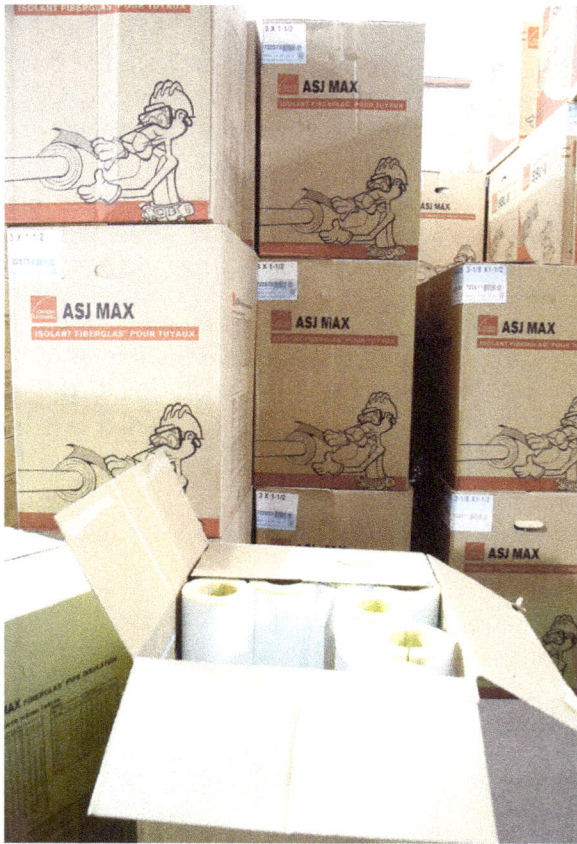

Figure 2 Open boxes should be stored at the front of the stack.

Figure 3 Materials stored on pallets and covered with a waterproof tarp.

• More than one cover might cause condensation beneath the covers, leading to moisture damage. Use one layer of waterproof covering only.

1.2.2 Storage Precautions

Some insulating materials require special storage precautions, such as the following:

• Items that are easily stolen, such as tape, wire, and staples, should be kept in secure (locked) storage. If storage place is not available, use a locked gang box.
• When ladders and scaffolds are not being used, they should be placed in secure storage.
• Store rolls of insulation protective metal jacket on their ends to prevent flat spots (*Figure 4*).
• Adhesive containers should be protected from very low or very high temperatures. They should be sealed tightly to prevent waste.
• Cellular foam insulation materials should be protected from direct sunlight.

Figure 4 Rolls of jacket should be stored on end.

- Aluminum or galvanized sheets can develop streaks or discoloration if they become damp. Keep them absolutely dry.
- Sheet metal is especially susceptible to damage from moisture and should not be stored outside for long periods. If sheet metal must be stored outside for a short time it must be completely covered with a tarp that is secured to the pallet. The tarp should be weighted in case of high wind.
- Welding pins should be stored in a dry area and not subjected to any moisture.

1.2.3 Maintaining Storage Areas

Like all work areas, the storage area should be kept as dry, clean, and orderly as is practicable. Use the following guidelines for maintaining storage areas:

- Stack materials by size and type.
- Remove waste materials such as empty boxes as soon as is reasonable. Pay attention to job site requirements for recycling and hazardous waste disposal.
- Return unused material to storage as soon as possible. If there is material in excess of the project needs, return that material to the supplier when possible.

Additional Resources

National Insulation Association (NIA) website offers resources for products and training, **www.insulation.org**

NCCER Module MT208, *Resource Control and Cost Awareness*.

1.0.0 Section Review

1. The packing list should contain _____.
   a. safety information for each item in a shipment
   b. a complete account of each item in a shipment
   c. the cost of each item in a shipment
   d. handling and storage cautions for each item in a shipment

2. Which one of the following statements is true about material storage areas?
   a. Materials should be placed according to when they will be needed.
   b. Materials should be placed according to their cost.
   c. Storage areas should be on upper floors, if possible, to avoid flooding.
   d. There should never be more than one storage area to make it easier to keep track of materials.

2.0.0 DISTRIBUTING INSULATION MATERIAL

Objective

Explain how to move and distribute insulating material.

a. Describe the methods used in moving insulation material.
b. Describe how to distribute insulating material.

Trade Terms

All-service jacketing (ASJ): A facing or covering applied to piping and mechanical equipment fiberglass insulation as a vapor barrier or for protection against abrasion.

Butt strips: Strips of ASJ that are several inches wide used to secure ends of piping insulation ASJ to each other after installation.

Construction elevator: An elevator dedicated to the movement of construction material and personnel.

HMIS label: A safety information label developed under the Federal Hazardous Materials Information System (HMIS) program that employers may apply to containers holding hazardous materials used in construction, industrial, government, and commercial activities.

NFPA fire diamond: A safety label established by the National Fire Protection Association (NFPA) used by emergency responders to understand the hazards of chemicals involved in a fire.

Rope and pulley: A hoisting system consisting of a rope running around grooved wheels which makes lifting heavy loads easier.

Staging: The placement of materials in smaller stacks close to the actual work site for easy access.

Once materials are received they are placed in one or more storage areas. They must be transported to the work area before they are needed by the mechanics.

2.1.0 Moving Insulation Material

As a trainee or apprentice, it is likely that you will be distributing materials to mechanics for installation. Remember that only insulation workers should be allowed to move insulation materials. Damaged materials will be the responsibility of the insulation distribution crew, even if they were damaged by someone else.

Materials are sometimes delivered by hand, one box or roll at a time. If the floor surface is smooth and there is sufficient room, you could also use a four-wheeled cart (*Figure 5*) for distribution, or simply slide several boxes along the floor.

On larger jobs and on jobs where storage is in a separate building, you might retrieve materials with a truck, forklift, or a utility tractor with a trailer. When using a truck or a trailer, make sure the loads are sufficiently secured for the trip, and pay attention to verify that no materials are shifting or loose enough to fall off the vehicle.

In facilities with multiple levels, you may have use of a construction elevator. Coordinate with other trades to schedule the use of the elevator. In existing facilities where you are doing maintenance and repair, verify that there is a freight elevator. If you have to use a regular elevator, make sure that protective blankets are hung inside the elevator to protect the elevator finishes.

> **WARNING!**
>
> Take the following safety precautions when moving materials to and from storage:
> - Wear high-visibility PPE to help ensure that you can be seen. Remember that work sites have numerous large or potentially dangerous vehicles, from heavy equipment to bulky material carts moved by persons with limited visibility. High-visibility PPE will help you stay safe.
> - Use proper lifting techniques. Lift with your legs and not your back, and use a lifting belt if the job calls for one.
> - Always barricade the work area when lifting insulation material by crane or overhead hoists.

2.1.1 Hoisting

Some large projects, such as power plant installations, need insulation materials on upper levels not served by a construction elevator. Materials bound for these levels must be hoisted. Be sure to use proper procedures and equipment, including slings, hooks, or boxes, to safely hoist materials. Small loads and loads not being lifted very far can be moved with a rope and pulley. Larger loads and loads being lifted higher might need a crane.

(A) FOUR-WHEELED CART

(B) FORKLIFT

Figure 5 Four wheeled cart and forklift.

2.2.0 Material Distribution

To get the right materials to the mechanics at the right time, you must first determine what materials the mechanics need. It is common for mechanics and crew members to pick up some items when heading to the work area in the morning, and again after lunch. Check with the mechanics in the morning and after lunch to verify that the materials at the work area are still needed. If not, they can be returned to storage.

Mechanics and crew installing insulating material often carry some items to a new work area, particularly material left over from a completed work area. Check with the crew to verify the materials that are needed. Make sure that anyone moving materials handles them with care to prevent damage.

After materials are delivered for the start of each shift, you need to know when more materials will be needed. Eventually, you will get a feel for the mechanic's speed and be able to predict when they might run out of material. If you are unsure of necessary distribution timing, check with the lead insulation mechanic to find out suggested timing for additional distribution.

As the saying goes, time is money. If an insulation mechanic is waiting for material to be delivered to his or her work area, the unproductive time is costing your company money. Try to stage your deliveries to get material to the mechanics before they are needed, and place the materials near where the mechanic will be installing them, as shown in *Figure 6*. Good staging procedures will keep work interruptions to a minimum, keep the mechanic productive, and make the project more cost-effective.

To avoid damage, only insulation workers should be allowed to move insulation materials. If materials need to be returned to storage, place them in the correct area. Separate the loose pieces by size and type, with one open box for each. If there is material in excess of the overall job needs, return it to storage, or if necessary to the distributor, as soon as practicable.

2.2.1 Placement of Materials for Safety

Materials placed in the work area should be protected from welding sparks. If upper levels are composed of steel grating, don't forget that a welder working at an upper level may create sparks which will fall through the grating to lower levels. Materials should also be placed away from areas where there will be open flames. When using any material with all-service jacketing (ASJ), leave the jacket butt strips in the box until they are needed. These strips are usually made of foil-faced Kraft paper and are easily damaged. Only one butt strip is provided for each section of insulation, so a damaged butt strip must be replaced at additional cost.

Figure 6 Materials staged for use by a mechanic.

When not in use, keep the lids on buckets or pails of mastics and adhesives tightly in place. Keep the material away from particularly hot or cold areas.

All insulation accessory items should be kept clean and protected from dust, water, and dampness when in storage.

If a hazardous insulation material (e.g., adhesives or mastics) is transferred from its original container to a container used at the work site, the new container must be labeled with the original safety information required by the Global Harmonization System (GHS).

The worksite superintendent may require applying an HMIS label (*Figure 7* [*A*]) for worker safety and an NFPA fire diamond (*Figure 7* [*B*]) for emergency responder safety that summarize this information to ready-use containers. This ensures that ready-use containers offer the same safety and fire protection information as the original container.

(A) HMIS LABEL

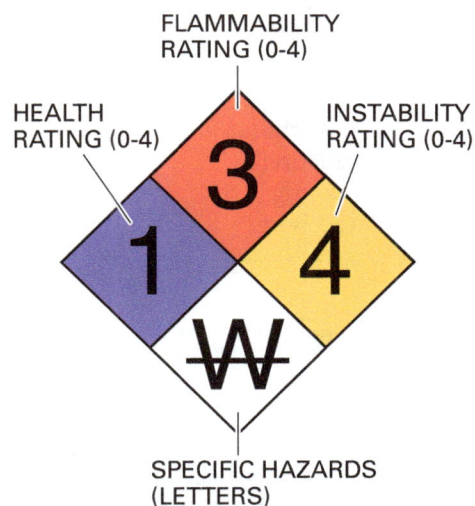

(B) NFPA FIRE DIAMOND

Figure 7 Sample HMIS and NFPA safety labels.

Hazardous Material Labeling in the Workplace

Safety labels on hazardous materials serve two purposes: They tell you what hazards a worker may face using the material and the kinds of PPE that should be worn when using it (HMIS labels). They also can identify the hazards, especially to emergency workers, resulting from the involvement of the material in a fire (NFPA fire diamond). The nine international GHS hazard class pictograms provide visual cues to the kinds of hazards present. They are required to be part of the original complete container hazard label that provides a lot of other technical information.

Users are permitted to summarize the chemical's safety information in HMIS and NFPA labels for ready-use containers. The numbers represent the level of hazard, from 0 (no hazard) to 4 (extreme hazard). Workers should note that the GHS numbering system is different. Its scale runs from 1 for extreme hazard to 5 for no hazard. You should receive training on the system of number scales and symbols being used in the safety labels at your job site as well as procedures for labeling ready use containers.

Additional Resources

National Insulation Association (NIA) website offers resources for products and training, **www.insulation.org**

NCCER Module MT208, *Resource Control and Cost Awareness*.

2.0.0 Section Review

1. When moving insulation materials to the next floor up, you should _____.
   a. ask a member of another construction trade for assistance
   b. use a construction elevator
   c. drag the packaged material up the stairs
   d. unpack and toss the materials to an assistant on the next floor

2. Ensuring that materials are delivered to a work area in easily accessible stacks for the mechanic is called _____.
   a. palletizing
   b. hoisting
   c. staging
   d. receiving

SUMMARY

Materials can amount to more than 50 percent of the cost of an insulating project. Some contractors even include handling and storage of materials as a separate line item on their bids. Because of potential impact on project cost, proper material handling is critical to the success and profitability of a project.

When a shipment of insulation material is delivered, the materials must be put into an appropriate storage area unless they are to be installed immediately. The storage area should be dry, clean, and secure to prevent damage or loss.

You must take similar care when distributing materials for installation. Getting materials to the mechanic where they are needed and before they are needed can help prevent downtime, which also affects the bottom line.

Review Questions

1. What should you do first when receiving a delivery of insulation materials?

   a. Move the materials to the site material storage area.
   b. Place the materials on pallets.
   c. Compare the packing slip or bill of lading to the materials delivered.
   d. Arrange the packages so that the labels are visible.

2. In order to sign for a shipment you must compare the items received to the _____.

   a. distribution staging
   b. original bid
   c. awarded contract
   d. packing slip

3. Delivered materials are typically _____.

   a. stacked in an appropriate storage area
   b. unloaded where the truck stops
   c. stacked in open air to prevent condensation
   d. stacked in work area where they will be used

4. A storage area must be _____.

   a. covered and dry
   b. enclosed and lockable
   c. beside the work area
   d. next to the receiving area

5. In insulation materials management, a *stack* refers to _____.

   a. layers of different types of insulation
   b. placement of like materials together
   c. an exhaust port from a furnace
   d. an assembly including equipment and insulation

6. A pallet refers to a _____.

   a. selection of insulation materials to give a mechanic options
   b. raised platform providing a work surface on an upper level
   c. support platform that keeps materials off the ground
   d. device used to raise and lower materials between levels

7. Construction materials that could be easily stolen should be _____.

   a. kept at home
   b. kept under armed guard
   c. kept in locked storage
   d. ordered as needed

8. Which of the following is recommended when using a construction elevator?

   a. Try to schedule use when the elevator is not being used by other crafts.
   b. Stage materials outside the elevator so they're ready to load as soon as the elevator is available.
   c. Use a construction elevator only when materials cannot be hoisted between floors.
   d. Use the construction elevator only for materials, never for crew.

9. As used with insulation storage and handling, *staging* refers to _____.

   a. the order in which insulation is applied
   b. the order in which materials arrive at the job site
   c. placement of materials on an elevated work platform
   d. placement of small stacks of materials next to work area

10. Which of the following may be a consideration when placing materials below levels that use grating as a floor?

   a. If anyone is eating above, food crumbs may fall through the grate and discolor the insulation.
   b. If material is to be used on the upper levels it will have to be moved again.
   c. If anyone is welding above, welding sparks may fall through the grate to the storage area.
   d. Workers on the grating may not be able to see the materials placed below them.

Trade Terms Quiz

Fill in the blank with the correct term that you learned from your study of this module.

1. _____ are used to secure ends of piping insulation ASJ to each other after installation.

2. _____ are metal framework used to support workers and materials for work on higher levels of a facility.

3. Every job site must have at least one _____ to hold insulation materials until they are needed.

4. Efficient _____ calls for material to be delivered to a work area just ahead of the worker who will be installing it.

5. All materials in storage should be on _____.

6. A(n) _____ is applied to hazardous materials mainly to provide workers with safety information when using the materials.

7. _____ is a covering of foil and Kraft paper used as a backing, primarily on fiberglass insulation.

8. Keep materials in storage in a(n) _____ with like materials and sizes together.

9. The _____ can be an easy way to move insulation materials to different floors, but you should coordinate its use with other craft professionals.

10. A(n) _____ should completely cover stored material before use to protect from weather and damage.

11. A(n) _____ are used to manually hoist materials up to a scaffold.

12. The mechanics and helpers assigned to a project are collectively called a(n) _____.

13. All shipments must have a(n) _____ that lists the contents of the shipment.

14. Emergency response personnel need to refer to the _____ label to understand how to deal with hazardous materials during a fire.

15. _____ may be used instead of a canvas tarpaulin to protect insulation materials from the weather and dust.

Trade Terms

All-service jacketing (ASJ)	HMIS label	Pallets	Stack
Butt strips	Material storage area	Polyethylene sheeting	Staging
Construction elevator	NFPA fire diamond	Rope and pulley	Tarpaulin
Crew	Packing slip	Scaffolds	

Trade Terms Introduced in This Module

All-service jacketing (ASJ): A facing or covering applied to piping and mechanical equipment fiberglass insulation as a vapor barrier or for protection against abrasion.

Butt strips: Strips of ASJ that are several inches wide used to secure ends of piping insulation ASJ to each other after installation.

Construction elevator: An elevator dedicated to the movement of material and personnel.

Crew: Mechanics and helpers assigned to a project. Each craft will have a crew for their specific applications.

HMIS label: A safety label required by the Federal Hazardous Materials Information System (HMIS) that is applied to all containers holding hazardous materials used in construction and industrial, government, and commercial activities.

Material storage area: A space located at the job site where all materials are stored.

NFPA fire diamond: A safety label established by the National Fire Protection Association (NFPA) used by emergency responders to understand the hazards of chemicals involved in a fire

Packing slip: A detailed listing of all materials shipped from a supplier. Every delivery/shipment has a packing slip.

Pallets: Small platforms used to place boxes on, to keep materials off the ground, and to make moving and delivering bulk materials easier.

Polyethylene sheeting: For storage purposes, heavy flexible sheets of plastic used as tarpaulins for protecting materials from the elements.

Rope and pulley: A hoisting system consisting of a rope running around grooved wheels which makes lifting heavy loads easier.

Stack: An orderly pile of like materials.

Staging: The placement of materials in smaller stacks close to the actual work site for easy access.

Scaffolds: Elevated work platforms for both personnel and materials.

Tarpaulin: Waterproofed fabric or canvas material that is used to cover insulation materials.

Additional Resources

This module presents thorough resources for task training. The following reference material is recommended for further study.

National Insulation Association (NIA) website offers resources for products and training, **www.insulation.org**
NCCER Module MT208, *Resource Control and Cost Awareness*.

Figure Credits

©iStockphoto.com/8thcreator, Module opener
Topaz Publications, Inc., Figure 2
Uline **www.Uline.com**, Figure 3
©iStockphoto.com/perkmeup, Figure 4
Gary Satterfield, Figure 5A
©iStockphoto.com/choja, Figure 5B
Luis Gonzalez, Master Qualifier of Powergas Services, Inc., Figure 6

Answer	Section Reference	Objective
Section One		
1. b	1.1.0	1a
2. a	1.2.1	1b
Section Two		
1. b	2.1.0	2a
2. c	2.2.0	2b

This page is intentionally left blank.

NCCER CURRICULA — USER UPDATE

NCCER makes every effort to keep its textbooks up-to-date and free of technical errors. We appreciate your help in this process. If you find an error, a typographical mistake, or an inaccuracy in NCCER's curricula, please fill out this form (or a photocopy), or complete the online form at **www.nccer.org/olf**. Be sure to include the exact module ID number, page number, a detailed description, and your recommended correction. Your input will be brought to the attention of the Authoring Team. Thank you for your assistance.

Instructors – If you have an idea for improving this textbook, or have found that additional materials were necessary to teach this module effectively, please let us know so that we may present your suggestions to the Authoring Team.

NCCER Product Development and Revision
13614 Progress Blvd., Alachua, FL 32615

Email: curriculum@nccer.org
Online: www.nccer.org/olf

❏ Trainee Guide   ❏ Lesson Plans   ❏ Exam   ❏ PowerPoints   Other _____

Craft / Level: _____ Copyright Date: _____

Module ID Number / Title: _____

Section Number(s): _____

Description: _____

_____

_____

_____

_____

Recommended Correction: _____

_____

_____

_____

_____

Your Name: _____

Address: _____

_____

Email: _____ Phone: _____

This page is intentionally left blank.

Characteristics of Pipe

OVERVIEW

Although insulation installers and mechanics do not install pipe systems, they routinely work around them and may be required to insulate them. The more you know about piping systems and the various types of piping and tubing, the easier it will be to plan your work.

Module 19105

Trainees with successful module completions may be eligible for credentialing through the NCCER Registry. To learn more, go to **www.nccer.org** or contact us at 1.888.622.3720. Our website has information on the latest product releases and training, as well as online versions of our *Cornerstone* magazine and Pearson's product catalog.

Your feedback is welcome. You may email your comments to **curriculum@nccer.org**, send general comments and inquiries to **info@nccer.org**, or fill in the User Update form at the back of this module.

This information is general in nature and intended for training purposes only. Actual performance of activities described in this manual requires compliance with all applicable operating, service, maintenance, and safety procedures under the direction of qualified personnel. References in this manual to patented or proprietary devices do not constitute a recommendation of their use.

19105 V2

Objectives

When you have completed this module, you will be able to do the following:

1. Describe pipe properties and construction.
   a. Describe pipe properties and sizing.
   b. Describe pipe construction.
2. Describe pipe system components.
   a. Describe pipe support components.
   b. Describe special-purpose pipe system components.

Performance Tasks

This is a knowledge-based module; there are no Performance Tasks.

Trade Terms

Butt weld
Copper tubing size (CTS)
Heat tracing
Inside diameter (ID)
Iron pipe size (IPS)
Nesting

Nominal
Outside diameter (OD)
Pipe insulation thickness
Pipe size
Pipe systems
Socket weld

Industry Recognized Credentials

If you are training through an NCCER-accredited sponsor, you may be eligible for credentials from NCCER's Registry. The ID number for this module is 19105. Note that this module may have been used in other NCCER curricula and may apply to other level completions. Contact NCCER's Registry at 888.622.3720 or go to **www.nccer.org** for more information.

Contents

Figures and Tables

1.0.0 PIPE PROPERTIES AND CONSTRUCTION

Objective

Describe pipe properties and construction.
  a. Describe pipe properties and sizing.
  b. Describe pipe construction.

Trade Terms

Butt weld: A pipe connection made by beveling the ends of two pieces of pipe and welding them together.

Copper tubing size (CTS): A pipe sizing standard. CTS pipe is characterized by thinner walls than iron pipe size.

Inside diameter (ID): The inside measurement of a section of insulation and/or the inside measurement of a section of pipe.

Iron pipe size (IPS): A pipe sizing standard. IPS is characterized by thicker walls than copper tubing size.

Nesting: Taking different sizes of like insulation and placing the smaller size inside the larger size to make the overall thickness greater.

Nominal: The target for a given measurement during manufacturing, such as the thickness of insulation. The material may actually be slightly less than or slightly greater than the nominal measurement.

Outside diameter (OD): The outside measurement of a section of insulation or a section of pipe.

Pipe insulation thickness: The thickness of insulation to be applied to the pipe system.

Pipe size: The measurement, expressed in inches, of the outside diameter (OD) and inside diameter (ID) of the pipe.

Pipe systems: Sections of pipe connected together to allow the movement of different processes.

Socket weld: A pipe connection made by sliding a collar over the two pieces of pipe to be joined and then welding the collar to the pipes.

When working with insulation, familiarity with pipe and pipe systems will help you work more efficiently. A working knowledge of pipe, tubing, connections, fittings, and valves can help you anticipate the best sequence for installation, which will facilitate scheduling of material deliveries and installation work.

1.1.0 Pipe Sizes

All pipe and tubing is identified by its size. The following four dimensions are used for this purpose:

- Outside diameter (OD) – The distance across the outer surface of the pipe. This is the surface where insulation is applied.
- Inside diameter (ID) – The distance across the inner surfaces of the pipe.
- *Wall thickness* – The distance between the inner and outer pipe surfaces.
- *Length* – The distance from one end of a pipe to the other. The most common length for pipe is 21', but 40' lengths are also common.

Pipe sizing nomenclature refers to the inner diameter. A pipe size is nominal, meaning that there is variance between the target size and the manufactured size of the pipe. As an insulator, the dimension that will be most important to you is the outside diameter; this is where you must fit insulation. Luckily, insulation is sized for pipe manufactured in each of the two common pipe sizing systems.

1.1.1 Steel and Copper Tubing Sizes

The inside diameter of pipe is of interest to plumbers and pipefitters, as well as architects and engineers. They must be aware of how much flow capacity the pipe offers. Mechanical insulators, on the other hand, are concerned with the outside diameter of the pipe. Pipes and tubing are sized in two different but related ways: iron pipe size (IPS) and copper tubing size (CTS). Each has its own variation in outside diameter.

IPS sizes define the nominal inside diameter of a pipe. For example, a 6" pipe measures 6" on the inside diameter. However, the outside diameter of a 6" pipe is $6\frac{5}{8}$". Insulation designed to fit a 6" IPS pipe is labeled *6" IPS*, and has an inside diameter of $6\frac{5}{8}$" (the same as the outside diameter of the pipe).

Like IPS, CTS is a nominal measure of inside diameter. However, because copper tubing walls are thinner, the outer diameter is smaller than a similarly named IPS pipe. For example, a copper tube with the same 6" nominal inside diameter will only measure 6⅛" at the outer diameter, which is much less than 6" IPS pipe. Insulation sized for the 6" copper tube is labeled *6" CTS*. Copper tubing sizes range from ½" CTS to 6" CTS.

Figure 1 shows a comparison of 6" IPS pipe and 6" CTS pipe.

> **NOTE**
>
> Pipe installers and mechanical insulators refer to the same pipe sizes differently, using either the inside diameter or outside diameter of the pipe. Refer to the *Appendix* at the end of this module for a table that illustrates this for various kinds of pipe.

Pipe and tubing are available in many types of materials, including carbon steel and stainless steel, various types of plastic (such as PVC), cast iron, glass, and fiberglass. Almost all pipe is sized according to its nominal IPS or CTS standards. Because insulation materials for pipes are available in sizes to fit both standards, there is piping insulation for almost any pipe available.

Some insulation sizes fit both IPS and CTS pipes. For example, ½" IPS insulation fits both ½" IPS pipe with an outside diameter of 0.840" and ¾" copper pipe with an outside diameter of 0.875". Other sizes of insulation, up to 1¼" IPS, can be used interchangeably on certain copper pipe sizes. This simplifies stocking requirements and

reduces costs, since distributors do not need to stock two different but almost identical products. Refer to *Table 1* for a schedule of interchangeable IPS and CTS sizes.

Schedules are one method of defining pipe wall thickness. A 6" Schedule 40 pipe has the same outer diameter as a 6" Schedule 80 pipe, as illustrated in *Figure 2*. The difference is the wall thickness. Since the Schedule 80 pipe has a great wall thickness, its inner diameter is smaller.

The same insulation fits all Schedules, since the change in wall thickness from one schedule to another affects only the pipe's inner diameter.

There are metric equivalents to IPS (*Table 2*). Metric pipes are sized according to the Diametre Nominal (DN) standard, which provides similar but not exact size equivalents to IPS sizes. In some cases, the difference in size may be so small that you can use IPS- or CTS-sized insulation, but other DN sizes may require a DN-sized insulation for proper fit.

DN sizes are used in Australia under the term NB, which stands for *Nominal Bore*. For example, a DN 25 mm pipe would be referred to as *25 NB* in Australia.

1.1.2 Pipe Insulation Thickness

Pipe insulation sizes are expressed in two dimensions: nominal pipe size and pipe insulation thickness. They are written as follows:

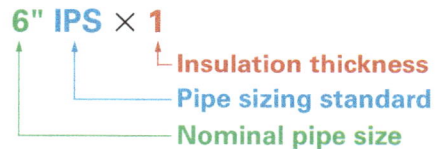

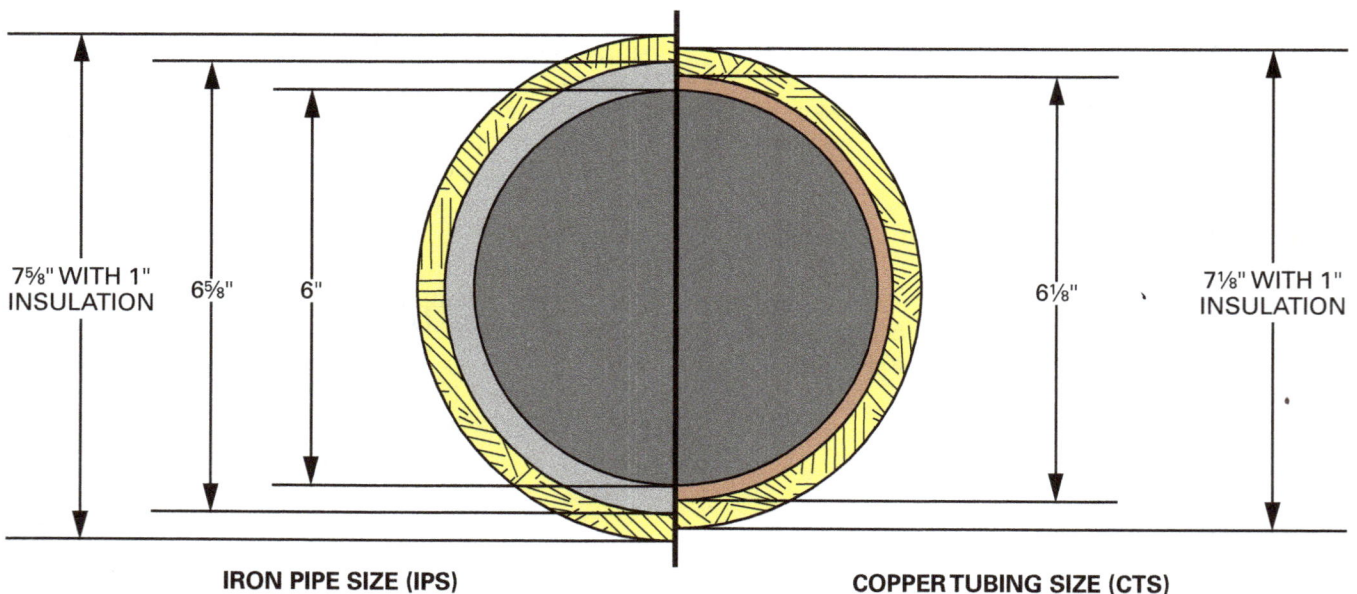

6" IPS × 1

— Insulation thickness
— Pipe sizing standard
— Nominal pipe size

Figure 1 6" IPS pipe and 6" CTS pipe compared.

7⅝" WITH 1" INSULATION   6⅝"   6"   **IRON PIPE SIZE (IPS)**

6⅛"   7⅛" WITH 1" INSULATION   **COPPER TUBING SIZE (CTS)**

Table 1 Interchangeable IPS and CTS Insulation Sizes

Nominal Pipe Size (ID)	Steel Pipe (IPS)	Copper Tubing (OD)	Steel Pipe (OD)	Interchangeable Steel Pipe (IPS)
½	½	⅝	⅞	None
¾	¾	⅞	1⅛	½
1	1	1⅛	1⅜	¾
1¼	1¼	1⅜	1⅝	1
1½	1½	1⅝	1⅞	1¼
2	2	2⅛	2⅜	None
2½	2½	2⅝	2⅞	None
3	3	3⅛	3½	None
4	4	4⅛	4½	None
5	5	5⅛	5⅝	None
6	6	6⅛	6⅝	None

This notation would call for insulation that is sized for 6" IPS pipe and is 1" thick.

As previously discussed, nominal pipe size is expressed as either an IPS or a CTS measurement. Similarly, insulation thickness is the nominal thickness of the insulating material. Note that nominal insulation thickness can vary, so a nominal 2" insulation product might be slightly more or less than 2" thick.

Project plans and specifications identify the thickness of the insulation to be applied to given equipment. While insulation is available in several thicknesses, the plan may call for a greater insulation thickness than is available. In such cases, insulation materials will be nested to achieve the required thickness. (Nesting refers to fitting larger insulation sizes over smaller ones to increase the overall thickness of the insulation.)

> **NOTE**
>
> Although insulation is manufactured to ASTM International recommended dimensions, there are allowable variations in dimensions due to manufacturing tolerances. This means that actual dimensions of materials may vary according to the type of material, the manufacturer, or even the batch. This may not generally be a concern, but when nesting a smaller product inside a larger product to achieve greater insulation thickness, tolerance variations may add up (or stack) to prevent proper fit.

1.2.0 Pipe Construction

Pipes commonly used in commercial and industrial applications are generally made from steel, copper, cast iron, or plastic. This section provides an overview of each type.

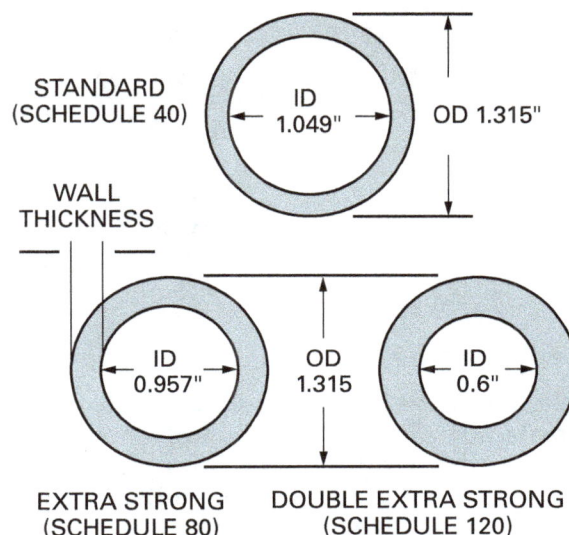

Figure 2 Steel pipe diameters.

Table 2 A Selection of DN Equivalents for IPS Pipes

Iron Pipe Size	Diametre Nominal Equivalent
¼"	8 mm
⅜"	10 mm
½"	15 mm
¾"	20 mm
1"	25 mm
1½"	40 mm
2"	50 mm
2½"	65 mm
3"	80 mm
4"	100 mm
6"	150 mm
8"	200 mm
10"	250 mm
12"	300 mm

Copper Tubing Types

Copper tubing is made with several wall thicknesses for different applications.

- Type M has slightly thicker walls and is generally used for residential and commercial water supply systems.
- Type L is slightly thicker walls than Type M. Like Type M, Type L is generally used for residential and commercial water systems.
- Type ACR, used for refrigerant piping in air conditioning and refrigeration, has the same wall thickness as Type L, but is sized by its outside diameter.
- Type K has the thickest walls. It is generally used in underground applications.

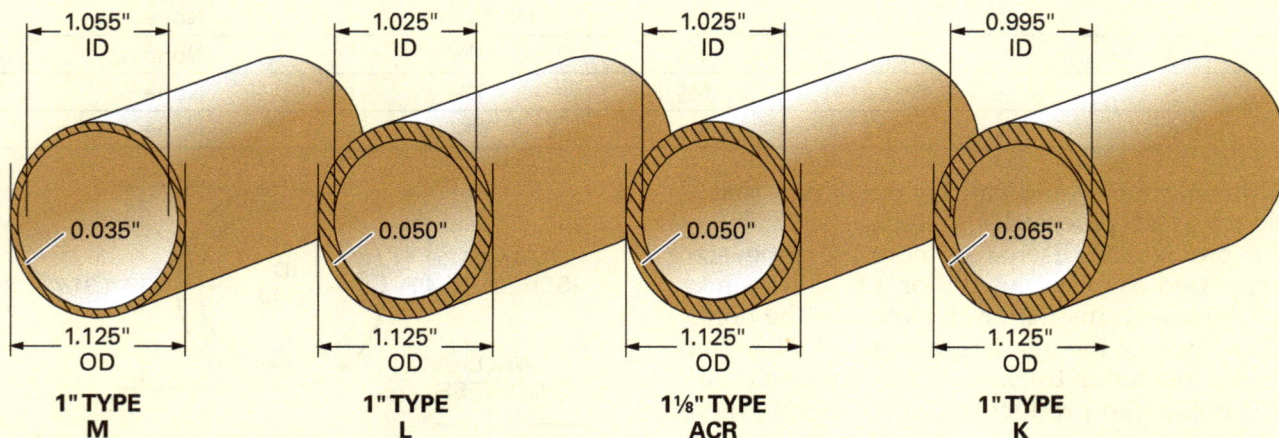

1.055" ID	1.025" ID	1.025" ID	0.995" ID
0.035"	0.050"	0.050"	0.065"
1.125" OD	1.125" OD	1.125" OD	1.125" OD
1" TYPE M	1" TYPE L	1⅛" TYPE ACR	1" TYPE K

1.2.1 Steel Pipe

Steel pipe (*Figure 3*) is made of carbon steel, which is created by mixing iron ore with carbon. Most steel pipe that requires insulation is sized in diameters ranging from ½" through 12".

Steel pipe is sized according to the IPS standard and is available in sizes starting at ½". Steel pipe sizes are available in different increments, depending on the size range. For example, steel pipe sizes are found in quarter-inch increments between ½" and 1½" IPS (½", ¾", 1", 1¼", and 1½"). The available increments of pipe sizes are shown in *Table 3*.

Standard steel pipe length is 21'. However, 40' lengths are also common. In commercial work, Schedule 40 pipe is generally used. On high-pressure or temperature systems, such as those found in power or industrial plants, heavier wall

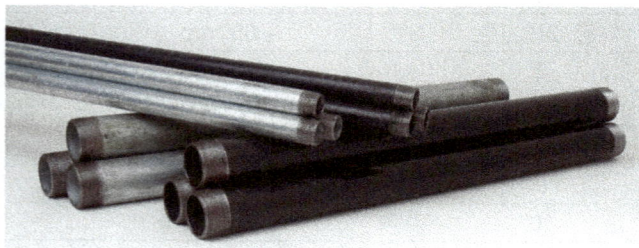

Figure 3 Steel pipe.

Table 3 Incremental Increases in Pipe Sizes

Pipe Size Range	Available Size Increments
≥ 1½"	¼" increments
1½" to 3"	½" increments
3" to 6"	1" increments
6" to 30"	2" increments

thickness pipe may be used. The heavier pipe (Schedule 80) does not affect the pipe size of the insulation since the outside pipe diameter remains the same.

There are three basic classifications of carbon steel pipe: black iron (sometimes called *black steel*), galvanized steel, and stainless steel. In industrial applications, most pipe systems requiring steel pipe are made from black iron. Black iron is not used in domestic water systems, however, due to the possibility of rust formation. Systems for domestic water use typically use galvanized steel or PVC.

Steel pipes may be connected to fittings, valves, specialty equipment, or other pipes. Steel pipe sections can be joined together using threaded fittings (*Figure 4* [A]), flanges (*Figure 4*[B]), or grooved fittings (*Figure 4*[C]). If connections are welded, pipes 2" and smaller will usually be connected with a butt weld (*Figure 5* [A]), while larger pipes and pipes that will see high-temperature or high-pressure duty are usually connected with a socket weld (*Figure 5* [*B*]).

If the pipe system design includes flanges, then they must also be insulated. Flanges can be significantly larger than the pipes they join, especially if the pipes are small. *Table 4* lists the dimensions for proper insulation fitment to insulate over flanges that operate within low pressure systems as recommended by ASTM International.

1.2.2 Cast Iron Pipe

Cast iron pipe is used in plumbing storm and sanitary drainage systems, including roof drains. There are two types of cast iron pipe: hub-and-spigot and no-hub (*Figure 6*). Cast iron pipe uses insulation sized for IPS pipes, although some sizes may fit loosely. Connections may be as follows:

- Hub-and-spigot type uses a gasket and compression fittings and is insulated by grooving out the straight pipe insulation.
- No-hub pipe uses stainless steel clamps with gaskets, and is insulated similarly to black steel flanges.

1.2.3 Copper Tubing

Copper tubing and piping (*Figure 7*) are used primarily on water, refrigeration, and some other systems where the operating temperatures do not exceed 200°F. Almost all refrigerant piping is copper.

Fittings such as those shown in *Figure 8* are used to connect lengths of copper piping or tubing. When used in industrial applications, copper is generally joined with brazed fittings. Brazing is a process in which a filler metal is melted with a heat in excess of 800°F (427°C) into joints between the fitting and the tubing or pipe.

Did You Know?

Galvanized pipe is carbon steel pipe that has been dipped in zinc. Galvanized pipe resists corrosion better than black iron. Stainless steel pipe is carbon steel pipe to which chromium has been added. Stainless steel pipe is used in corrosive environments and in processing food, pharmaceuticals, and other products where contamination must be prevented.

(A) THREADED FITTING

(B) FLANGE

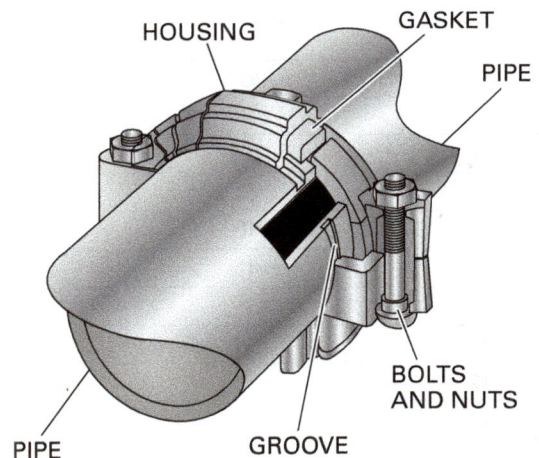

(C) GROOVED FITTING

Figure 4 Threaded, flanged, and grooved fittings.

(A) BUTT WELD

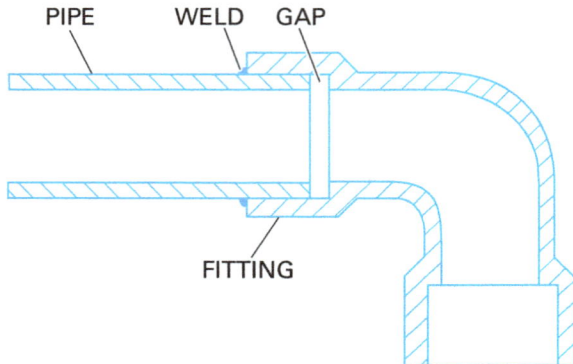

(B) SOCKET WELD

Figure 5 Butt-weld and socket-weld pipe joints.

Some copper tubing, especially tubing that is used in lower pressure applications, may be joined by soldering. Soldering is similar to brazing but uses a much lower temperature to melt a flexible filler metal.

Compression-type connections are occasionally used, particularly where short sections of pipe are connected to equipment. Brazed and soldered connections are insulated by grooving-out or compressing the straight pipe insulation.

1.2.4 Plastic Pipe

Plastic pipe (*Figure 9*) is used in a variety of residential, commercial, and industrial applications. Plastic pipe is manufactured from several types of plastic, including the following:

- Polyvinyl chloride (PVC)
- Chlorinated polyvinyl chloride (CPVC)
- Acrylonitrile Butadiene Styrene (ABS)
- Polyethylene (PE)
- Cross-linked Polyethylene (PEX)

The majority of plastic pipe in commercial and industrial applications is made from PVC.

Table 4   Dimensions of Insulation for Line Flanges

Iron Pipe Size	OD of Flange	Insulation Size	"A" Thickness Flanges	"B" Inside Allowing for Bolts
$1\frac{1}{2}$	$2\frac{1}{2}$	3	$1\frac{1}{2}$	$2\frac{1}{2}$
$\frac{3}{4}$	3	4	2	3
1	$4\frac{1}{4}$	4	$2\frac{1}{4}$	$3\frac{1}{4}$
$1\frac{1}{4}$	4	5	$2\frac{1}{2}$	$3\frac{1}{2}$
$1\frac{1}{2}$	5	5	$2\frac{3}{4}$	$3\frac{3}{4}$
2	6	6	3	4
$2\frac{1}{2}$	7	7	$3\frac{1}{2}$	$4\frac{1}{2}$
3	$7\frac{1}{2}$	7	$3\frac{3}{4}$	$5\frac{1}{4}$
4	9	9	$3\frac{3}{4}$	$5\frac{1}{4}$
5	10	10	$3\frac{3}{4}$	$5\frac{1}{4}$
6	11	11	4	$5\frac{1}{2}$
8	$13\frac{1}{2}$	13	$4\frac{1}{2}$	6
10	16	16	$4\frac{3}{4}$	$6\frac{1}{4}$
12	19	19	5	$6\frac{3}{4}$
14	21	21	$5\frac{1}{2}$	7
16	$23\frac{1}{2}$	24	$5\frac{3}{4}$	$7\frac{1}{4}$
18	25	25	$6\frac{1}{4}$	$7\frac{3}{4}$
20	$27\frac{1}{2}$	28	$6\frac{3}{4}$	$8\frac{1}{4}$
24	32	32	$6\frac{3}{4}$	$8\frac{1}{4}$

(A) NO-HUB

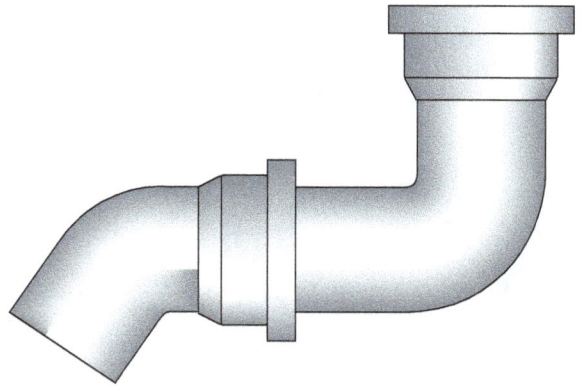

(B) HUB-AND-SPIGOT

Figure 6 Cast iron pipe.

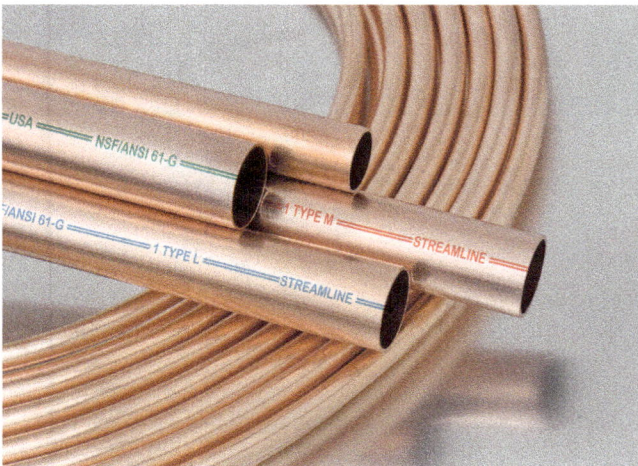

Figure 7 Copper pipe and tubing.

Plastic pipe is used in plumbing water and waste, industrial processes, and other systems where the operating temperatures are less than 200°F. Pipe sizes and ODs for plastic pipe are the same as for steel pipe. (Plastic pipe uses the IPS sizing standard.) Plastic pipe connectors and fittings are the same configuration as the screwed or socket-weld connectors for steel pipe. The insulation treatment at the connections and fittings is the same as for threaded steel pipe.

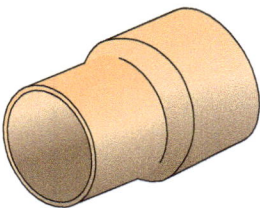

REDUCING BUSHING

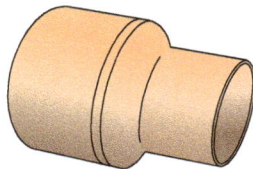

REDUCING COUPLING

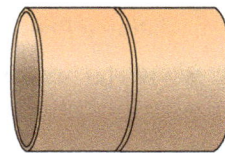

COUPLING

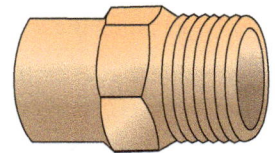

MALE ADAPTER

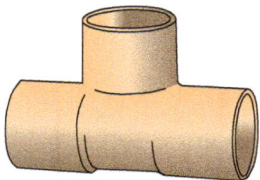

TEE

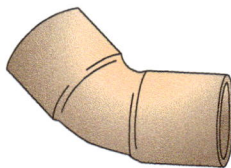

45° ELBOW

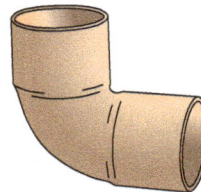

**SHORT RADIUS
90° ELBOW**

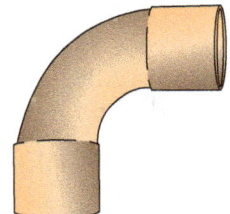

**LONG RADIUS
90° ELL**

Figure 8 Copper pipe fittings.

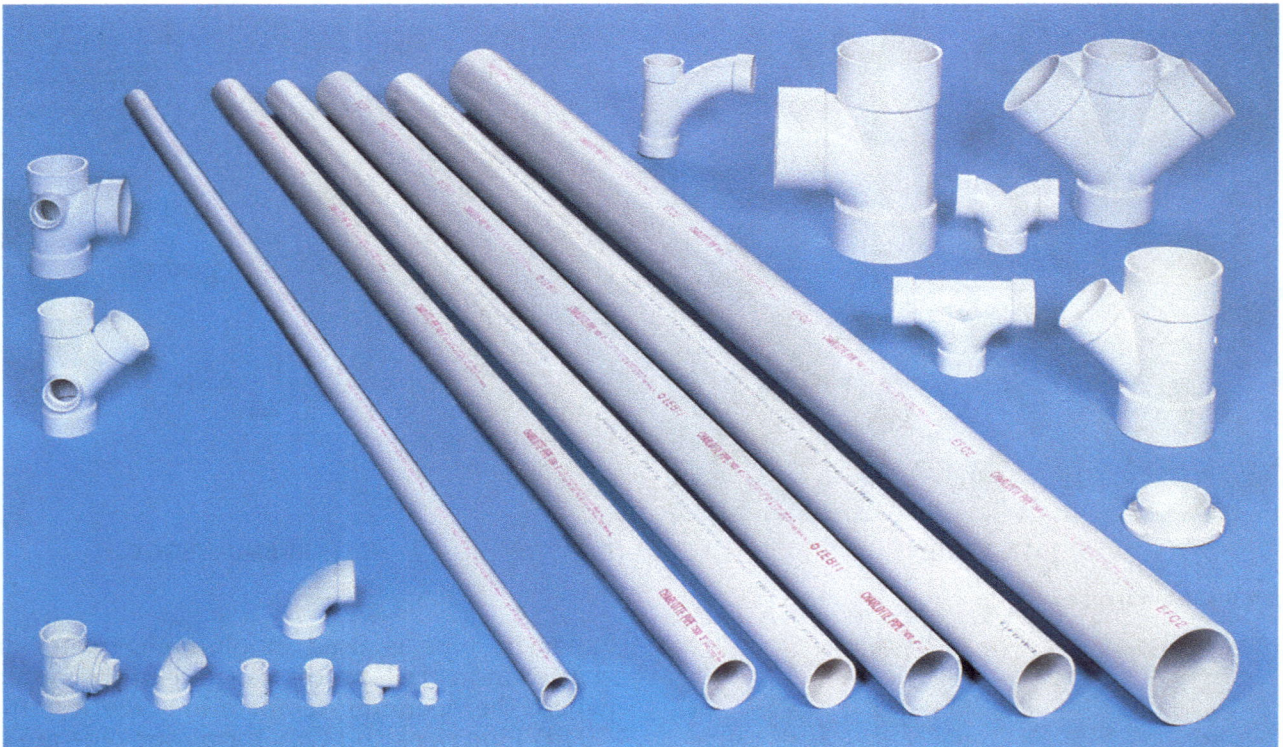

Figure 9 PVC pipe and fittings.

Did You Know?

The terms *tube* and *pipe* are often used interchangeably, but there are slight differences between them. Experienced technicians and installers generally consider tubes to have relatively thin walls with some level of flexibility, and pipe to have relatively thick walls.

Additional Resources

The following websites offer resources for products and training:

Plastic Pipe and Fittings Association (PPFA), **www.ppfahome.org**

Plastics Pipe Institute (PPI), **www.plasticpipe.org**

1.0.0 Section Review

1. Copper tubing is sized according to which standard?
   a. IPS
   b. CTS
   c. ABS
   d. PEX

2. Which of the following is a type of plastic pipe?
   a. PVC
   b. OD
   c. ID
   d. Type M

2.0.0 PIPE SYSTEM COMPONENTS

Objective

Describe pipe system components.
   a. Describe pipe support components.
   b. Describe special-purpose pipe system
       components.

Trade Term

Heat tracing: A form of applying heat to the outside surface of a pipe to keep the internal product at a certain temperature.

While pipe makes up the bulk of most plumbing systems, there are many other components. Some of these components must also be insulated. In addition, the insulation mechanic must plan installation to work with other components even if they do not require insulation. These components include hangers, anchors, expansion provisions, valves, and heat tracing systems.

2.1.0 Pipe Support Components

In order to support the combined weight of piping runs and the materials that flow through them, the piping must be secured to overheads and vertical surfaces. Pipe hangers, supports, and anchors are used for this purpose.

2.1.1 Hangers and Supports

Generally, cold piping systems have pipe hangers sized to allow the pipe insulation to pass through the hanger. This allows for insulation and a vapor barrier that are both continuous, without gaps or interruptions, which will increase the possibility of condensation. To help prevent the insulation from being compressed, high-density foam inserts are sometimes used to spread the load across the insulation's surface. There are many types of pipe hangers. *Figure 10* shows some examples.

Hot piping systems sometimes use the same type of hanger, often in direct contact with the pipe. Many hot pipe systems have steel saddles welded directly to the pipe, supported by standard clevis hangers. Because high-temperature and high-pressure systems often experience movement of the pipes as temperatures and

pressures fluctuate, these systems frequently use roller hangers (*Figure 11*).

2.1.2 Pipe Anchors

Pipe anchors are located along a pipe run to stabilize the pipe and control the direction of pipe expansion. There are numerous ways to secure piping to a surface. Examples are shown in *Figure 12*. Anchors may be bolted, welded, or clamped to the pipe and connected to an immovable part of the structure.

Pipe anchors used in low-temperature applications must be insulated. As a rule of thumb, the insulation used on the anchor should cover a distance equal to three times the thickness of the pipe insulation. This helps prevent condensation on the anchor.

2.2.0 Special-Purpose Pipe Components

Other pipe system components perform special functions, such as allowing for thermal expansion and contraction in the system, or enabling material flow inside the system.

2.2.1 Pipe Expansion Components

Material expands when it is heated, and contracts when it is cooled. Engineers are forced to account for thermal expansion and contraction in several systems, including highway bridges and automotive engine components. Engineers designing pipe systems that carry hot or cold materials must also account the thermal expansion and contraction induced by those materials.

Expansion loops and expansion joints (*Figure 13*) are pipe system components that allow for thermal expansion and contraction.

Expansion loops are structures made of regular pipe materials. They are typically U-shaped.

Bellows expansion joints are inline components with integrated sections of rubber or pleated metal, both of which can withstand thermal expansion and contraction. *Figure 14* shows a close-up view of a bellows expansion joint.

Table 5 shows the amount of linear expansion (expressed in inches per 100 linear feet of pipe) for different types of pipe at various temperatures.

2.2.2 Heat Tracing Systems

For some pipe systems carrying heated material, maintaining the material's thermal conditioning is critical. These pipe systems use heat tracing to help maintain the temperature of the pipe's contents. In addition, the thermal insulation on

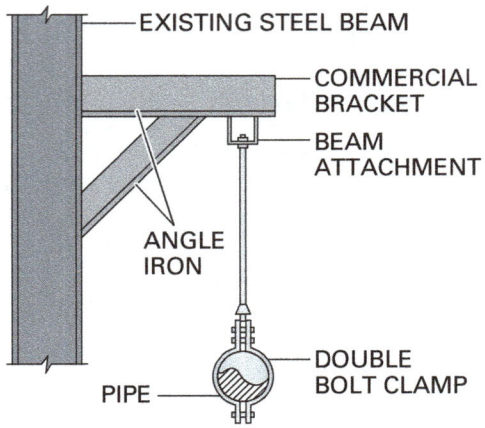

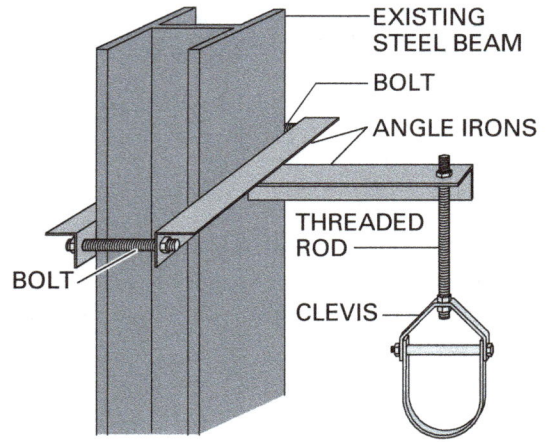

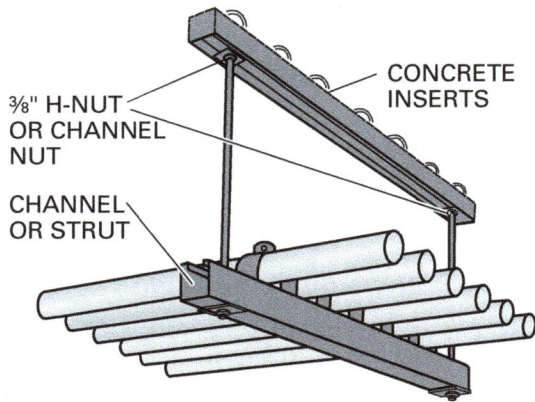

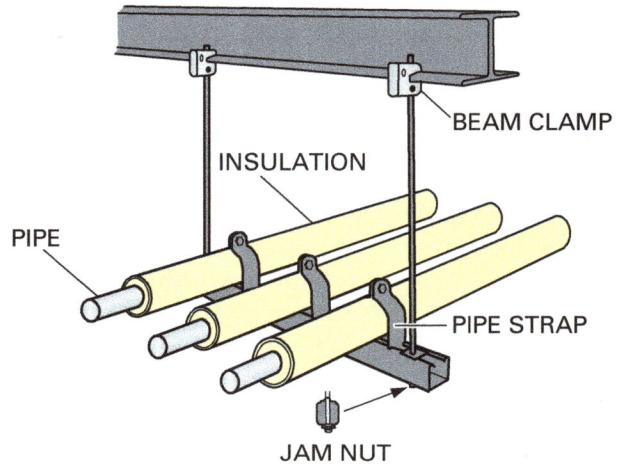

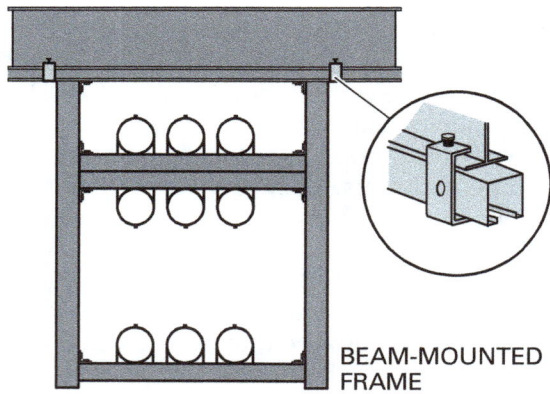

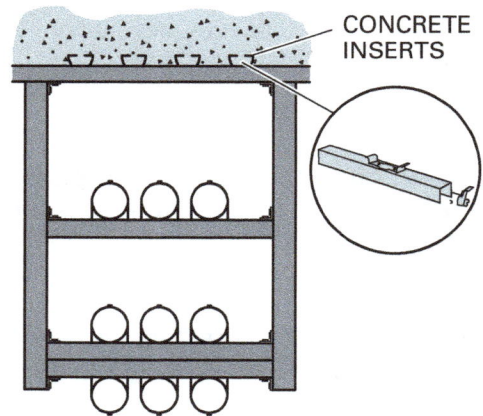

Figure 10 Pipe hangers.

Figure 11 Pipe roller support.

pipes exposed to very cold climates may not be adequate to prevent the process fluid from freezing. In such cases, heat tracing systems are used to maintain the temperature.

Heat tracing systems may use hot fluid, steam, or electricity to provide heat. Steam systems use small tubing wrapped around the pipe to carry steam. Steam systems are being increasingly replaced by electrical systems, which wrap the pipe with electric cable (*Figure 15*) or tape which generates heat when current is passed through it. Other systems use tubing that carries hot oil along the pipe.

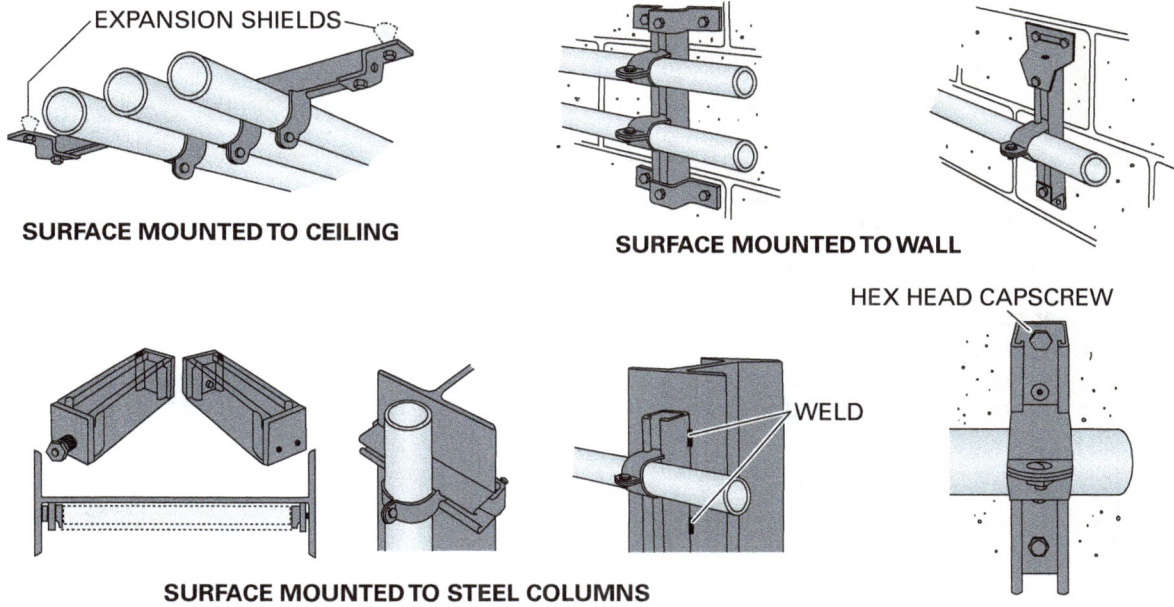

SURFACE MOUNTED TO CEILING

SURFACE MOUNTED TO WALL

SURFACE MOUNTED TO STEEL COLUMNS

Figure 12 Pipe anchors.

Did You Know?

In areas that are subject to earthquakes, special arrangements are used for pipe supports. As shown here, a spring is included in the pipe hanger. The two diagonal cables are used to limit sideways movement.

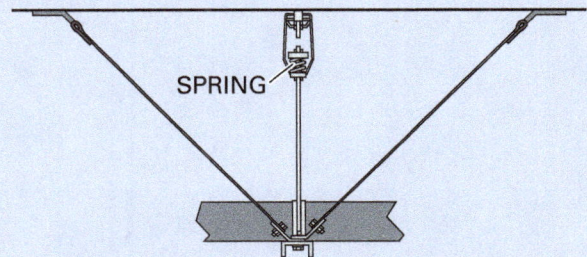

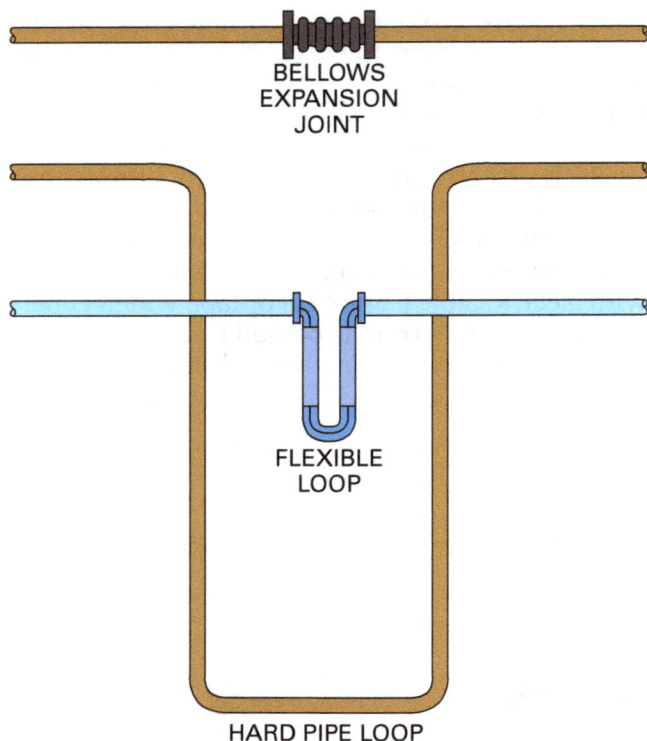

Figure 13 Expansion joint and expansion loop.

Any of the heat tracing systems can be applied in straight runs or wrapped around a pipe in a spiral. Either system effectively increases the outer diameter of the pipe on which it is installed.

Figure 14 Bellows expansion joint.

Be sure that the insulation design specifications you're given take the added space required into account when specifying an insulation to be applied to the pipe.

2.2.3 Valves and Specialty Components

Piping systems generally contain valves, and may also include specialty components such as regulators, strainers, orifices, and other devices. Insulating these items requires skill and careful consideration. Be sure you know the functions and service requirements of these specialty components before proceeding with insulation work involving them. Figure 16 shows several types of valves.

Table 5 Thermal Expansion of Pipes in Inches Per 100 Linear Feet

Temperature	Cast Iron Pipe	Steel Pipe	Wrought Iron Pipe	Copper Pipe	Temperature	Cast Iron Pipe	Steel Pipe	Wrought Iron Pipe	Copper Pipe
−20°F	0	0	0	0	500°F	3.847	4.296	4.477	6.110
0°F	0.127	0.145	0.152	0.204	520°F	4.020	4.487	4.466	6.352
20°F	0.255	0.293	0.306	0.442	540°F	4.190	4.670	4.866	6.614
40°F	0.390	0.430	0.465	0.655	560°F	4.365	4.860	5.057	6.850
60°F	0.518	0.593	0.620	0.888	580°F	4.541	5.051	5.268	7.123
80°F	0.649	0.725	0.780	1.100	600°F	4.725	5.247	5.455	7.388
100°F	0.787	0.898	0.939	1.338	620°F	4.896	5.437	5.660	7.636
120°F	0.926	1.055	1.110	1.570	640°F	5.082	5.627	5.850	7.893
140°F	1.051	1.209	1.265	1.794	660°F	5.260	5.831	6.067	8.153
160°F	1.200	1.368	1.427	2.008	680°F	5.442	6.030	6.260	8.400
180°F	1.345	1.528	1.597	2.255	700°F	5.629	6.229	6.481	8.676
200°F	1.495	1.691	1.778	2.500	720°F	5.808	6.425	6.673	8.912
220°F	1.634	1.852	1.936	2.720	740°F	6.006	6.635	6.899	9.203
240°F	1.780	2.020	2.110	2.960	760°F	6.200	6.833	7.100	9.460
260°F	1.931	2.183	2.279	3.189	780°F	6.389	7.046	7.314	9.736
280°F	2.085	2.350	2.465	3.422	800°F	6.587	7.250	7.508	9.992
300°F	2.233	2.519	2.630	3.665	820°F	6.779	7.464	7.757	10.272
320°F	2.395	2.690	2.800	3.900	840°F	6.970	7.662	7.952	10.512
340°F	2.543	2.862	2.988	4.145	860°F	7.176	7.888	8.195	10.814
360°F	2.700	3.029	3.175	4.380	880°F	7.375	8.098	8.400	11.175
380°F	2.859	3.211	3.350	4.628	900°F	7.579	8.313	8.639	11.360
400°F	3.008	3.375	3.521	4.870	920°F	7.795	8.545	8.867	11.625
420°F	3.182	3.566	3.720	5.118	940°F	7.898	8.755	9.089	11.911
440°F	3.345	3.740	3.900	5.358	960°F	8.200	8.975	9.300	12.180
460°F	3.511	3.929	4.096	5.612	980°F	8.406	9.196	9.547	12.473
480°F	3.683	4.100	4.280	5.855	1,000°F	8.617	9.421	9.776	12.747

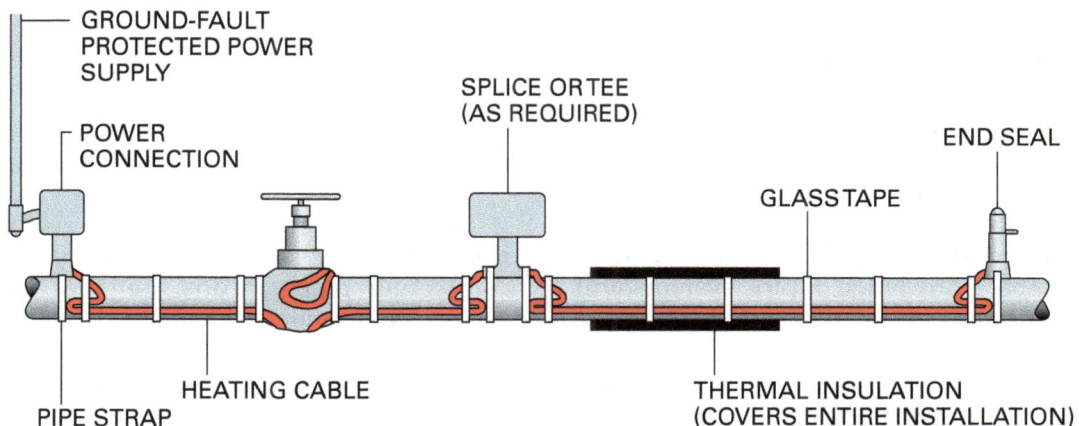

Figure 15 Electrical cable heat tracing.

(A) GATE VALVE **(B) PRESSURE REGULATOR VALVE** **(C) BUTTERFLY CHECK VALVE**

Image (B) courtesy of Watts Regulator Company

Figure 16 Examples of valves.

2.0.0 Section Review

1. Which one of the following allows for the movement of pipe due to fluctuations in temperature and pressure?

    a. Heat tracers
    b. Anchors
    c. Clevis hangers
    d. Roller supports

2. Which of the following is a type of heat tracing system?

    a. Steam
    b. Solar
    c. Friction
    d. Hot polymer

SUMMARY

Insulation mechanics do not install pipe, and trainees may wonder why they would need to know about pipe. Knowledge of pipe systems will make you a far more effective and efficient insulation mechanic.

Pipe is generally available in two sizing standards: IPS and CTS. For a given size, pipe in each standard will have the same nominal inner diameter, but the wall thicknesses, and therefore the outer diameters, will vary. As an insulation mechanic, you are concerned with fitting insulation to the outer surface, so you'll need to know which sizing standard is being used for the equipment you need to insulate.

Pipes are available in several different materials. In addition, there are several methods of connecting pipes to one another, and to other system components. Plastic components (including PVC) are frequently glued, for example, while iron components are either welded together or machined for threading.

There are several pipe system components which will also require insulation. Some of these components allow for the expansion and contraction of pipe due to heat. Other components are used to enable material flow, or to maintain a material's thermal conditioning.

NCCER – *Mechanical Insulating*

1. Which of the following pipe measurements is of the most interest to an insulator?

   a. inside diameter
   b. outside diameter
   c. wall thickness
   d. overall length

2. CTS is an acronym for _____.

   a. coupling tubing size
   b. compression tubing size
   c. copper tubing size
   d. continuous tubing size

3. Pipe insulation sizes are expressed in two dimensions: nominal pipe size and _____.

   a. pipe insulation thickness
   b. inner piping diameter
   c. pipe wall thickness
   d. outer piping diameter

4. The procedure used when one layer of insulation will not meet the specification for insulation thickness is called _____.

   a. layering
   b. gluing
   c. nesting
   d. foaming

5. Which of the following is used for joining steel pipe?

   a. Glue
   b. Threaded fittings
   c. Brazing
   d. Cup and socket

6. What is the most common type of plastic pipe in commercial and industrial applications?

   a. ABS
   b. PEX
   c. PVC
   d. CPVC

7. Plastic pipe is used where _____.

   a. pipes are subject to freezing
   b. operating temperatures are less than 200°F (93°C)
   c. operating temperatures 250°F (121°C) or more
   d. excessive material pressure stresses metal pipe fasteners

8. Which of the following is used to prevent insulation from being compressed by the weight of the pipe it covers?

   a. Clevis hangers
   b. High-density inserts
   c. Vermiculite
   d. All service jacket

9. An expansion loop is a structure that _____.

   a. allows for thermal expansion and contraction
   b. allows for pipe stretch when hanging vertically
   c. transitions from one pipe size to another
   d. absorbs the force of accelerated material in a pipe system

10. Heat tracing is _____.

    a. the effect of heated material in a pipe on that pipe's connections
    b. the slow thinning of pipe walls due to carrying heating material
    c. a method of diagnosing excessive heat loss
    d. a method of maintaining material temperature

Fill in the blank with the correct term that you learned from your study of this module.

1. The distance across the inner surfaces of a pipe is called the _____.

2. Installing one layer of insulation over another layer to increase the overall thickness is called _____.

3. Using a system to apply heat to the outside of a pipe to minimize heat loss is called _____.

4. The distance from the inner surface of a section of applied insulation to the outer surface is called the _____.

5. The nominal description of pipe dimensions is called the _____.

6. Sections of pipe or tubing that, when connected together, allow for the contained movement of materials are called _____.

7. The larger of the two pipe sizing standards is the _____.

8. When two pieces of pipe are joined by beveling and welding the ends together, the result is called a(n) _____.

9. The smaller of the two pipe sizing standards is the _____.

10. When a collar is slipped across the connection of two pipes and welded in place on both the result is called a(n) _____.

11. The distance across the outer surfaces of a pipe is called the _____.

12. A dimension that varies based on other dimensions or inputs is said to be _____.

Trade Terms

Butt weld
Copper tubing size (CTS)
Heat tracing
Inside diameter (ID)

Iron pipe size (IPS)
Nesting
Nominal
Outside diameter (OD)

Pipe insulation thickness
Pipe size
Pipe systems
Socket weld

Appendix

Pipe and Insulation Chart

Pipe ID (inches) Copper	Pipe OD (inches)	Insulator Terminology	Pipe Installer Terminology
$\frac{1}{2}$	$\frac{5}{8}$	$\frac{5}{8}$	$\frac{1}{2}$
$\frac{3}{4}$	$\frac{7}{8}$	$\frac{7}{8}$	$\frac{3}{4}$
1	$1\frac{1}{8}$	$1\frac{1}{8}$	1
$1\frac{1}{4}$	$1\frac{3}{8}$	$1\frac{3}{8}$	$1\frac{1}{4}$
$1\frac{1}{2}$	$1\frac{5}{8}$	$1\frac{5}{8}$	$1\frac{1}{2}$
2	$2\frac{1}{8}$	$2\frac{1}{8}$	2
$2\frac{1}{2}$	$2\frac{5}{8}$	$2\frac{5}{8}$	$2\frac{1}{2}$
3	$3\frac{1}{8}$	$3\frac{1}{8}$	3
$3\frac{1}{2}$	$3\frac{5}{8}$	$3\frac{5}{8}$	$3\frac{1}{2}$
4	$4\frac{1}{8}$	$4\frac{1}{8}$	4
5	$5\frac{1}{8}$	$5\frac{1}{8}$	5
6	$6\frac{1}{8}$	$6\frac{1}{8}$	6
Iron (IPS) and Stainless Steel			
$\frac{1}{2}$	$\frac{7}{8}$	$\frac{1}{2}$	$\frac{1}{2}$
$\frac{3}{4}$	$1\frac{1}{8}$	$\frac{3}{4}$	$\frac{3}{4}$
1	$1\frac{3}{8}$	1	1
$1\frac{1}{4}$	$1\frac{5}{8}$	$1\frac{1}{4}$	$1\frac{1}{4}$
$1\frac{1}{2}$	$1\frac{7}{8}$	$1\frac{1}{2}$	$1\frac{1}{2}$
2	$2\frac{3}{8}$	2	2
$2\frac{1}{2}$	$2\frac{7}{8}$	$2\frac{1}{2}$	$2\frac{1}{2}$
3	$3\frac{1}{2}$	3	3
$3\frac{1}{2}$	4	$3\frac{1}{2}$	$3\frac{1}{2}$
4	$4\frac{1}{2}$	4	4
5	$5\frac{1}{2}$	5	5
6	$6\frac{5}{8}$	6	6
8	$8\frac{5}{8}$	8	8
10	$10\frac{3}{4}$	10	10
12	$12\frac{3}{4}$	12	12
Stainless Steel Tubing			
All measurements for ID are in millimeters and are not listed here.	1	1	1
	$1\frac{1}{2}$	$1\frac{1}{2}$	$1\frac{1}{2}$
	2	2	2
	$2\frac{1}{2}$	$2\frac{1}{2}$	$2\frac{1}{2}$
	3	3	3
	$3\frac{1}{2}$	$3\frac{1}{2}$	$3\frac{1}{2}$
	4	4	4

Pipe ID (inches) PVC Pipe Schedule 40 and 80	Pipe OD (inches)	Insulator Terminology	Pipe Installer Terminology
$\frac{1}{2}$	$\frac{7}{8}$	$\frac{1}{2}$	$\frac{1}{2}$
$\frac{3}{4}$	$1\frac{1}{8}$	$\frac{3}{4}$	$\frac{3}{4}$
1	$1\frac{3}{8}$	1	1
$1\frac{1}{4}$	$1\frac{5}{8}$	$1\frac{1}{4}$	$1\frac{1}{4}$
$1\frac{1}{2}$	$1\frac{7}{8}$	$1\frac{1}{2}$	$1\frac{1}{2}$
2	$2\frac{3}{8}$	2	2
$2\frac{1}{2}$	$2\frac{7}{8}$	$2\frac{1}{2}$	$2\frac{1}{2}$
3	$3\frac{3}{8}$	3	3
$3\frac{1}{2}$	$4\frac{1}{8}$	$3\frac{1}{2}$	$3\frac{1}{2}$
4	$4\frac{5}{8}$	4	4
5	$5\frac{5}{8}$	5	5
6	$6\frac{5}{8}$	6	6
8	$8\frac{5}{8}$	8	8
10	$10\frac{3}{4}$	10	10
12	$12\frac{3}{4}$	12	12

NCCER – *Mechanical Insulating*

Trade Terms Introduced in This Module

Butt weld: A pipe connection made by beveling the ends of two pieces of pipe and welding them together.

Copper tubing size (CTS): A pipe sizing standard. CTS pipe is characterized by thinner walls than iron pipe size.

Heat tracing: A form of applying heat to the outside surface of a pipe to keep the internal product at a certain temperature.

Inside diameter (ID): The inside measurement of a section of insulation and/or the inside measurement of a section of pipe.

Iron pipe size (IPS): A pipe sizing standard. IPS is characterized by thicker walls than copper tubing size.

Nesting: Taking different sizes of like insulation and placing the smaller size inside the larger size to make the overall thickness greater.

Nominal: The target for a given measurement during manufacturing, such as the thickness of insulation. The material may actually be slightly less than or slightly greater than the nominal measurement.

Outside diameter (OD): The outside measurement of a section of insulation or a section of pipe.

Pipe insulation thickness: The thickness of insulation to be applied to the pipe system.

Pipe size: The measurement, expressed in inches, of the outside diameter (OD) and inside diameter (ID) of the pipe.

Pipe systems: Sections of pipe connected together to allow the movement of different processes.

Socket weld: A pipe connection made by sliding a collar over the two pieces of pipe to be joined and then welding the collar to the pipes.

Additional Resources

This module presents thorough resources for task training. The following reference material is recommended for further study.

The following websites offer resources for products and training:
Plastic Pipe and Fittings Association (PPFA), **www.ppfahome.org**
Plastics Pipe Institute (PPI), **www.plasticpipe.org**

Figure Credits

Courtesy of NOV Fiber Glass Systems, Mod Opener
Mueller Streamline Co., Figures 3, 4A, 7
Apollo Valves and Elkhart Products Corporation, Figures 4B, 16A, 16C
Topaz Publications, Inc., Figure 5A
Courtesy of Charlotte Pipe and Foundry, Figures 6A, 9
Anvil International, LLC, Exeter, NH, Figure 11
P D Blowers, Figure 14
Pentair Thermal Management, Figure 15
Courtesy of Watts Regulator Company, Figure 16B
Peter J. Gauchel, Appendix

Answer	Section Reference	Objective
Section One		
1. b	1.1.1	1a
2. a	1.2.4	1b
Section Two		
1. d	2.1.1	2a
2. a	2.2.2	2b

This page is intentionally left blank.

NCCER CURRICULA — USER UPDATE

NCCER makes every effort to keep its textbooks up-to-date and free of technical errors. We appreciate your help in this process. If you find an error, a typographical mistake, or an inaccuracy in NCCER's curricula, please fill out this form (or a photocopy), or complete the online form at **www.nccer.org/olf**. Be sure to include the exact module ID number, page number, a detailed description, and your recommended correction. Your input will be brought to the attention of the Authoring Team. Thank you for your assistance.

Instructors – If you have an idea for improving this textbook, or have found that additional materials were necessary to teach this module effectively, please let us know so that we may present your suggestions to the Authoring Team.

NCCER Product Development and Revision

13614 Progress Blvd., Alachua, FL 32615

Email: curriculum@nccer.org
Online: www.nccer.org/olf

❏ Trainee Guide   ❏ Lesson Plans   ❏ Exam   ❏ PowerPoints   Other _____

Craft / Level: _____ Copyright Date: _____

Module ID Number / Title: _____

Section Number(s): _____

Description: _____

_____

_____

_____

_____

Recommended Correction: _____

_____

_____

_____

_____

Your Name: _____

Address: _____

_____

Email: _____ Phone: _____

This page is intentionally left blank.

Plumbing Systems

OVERVIEW

Insulation mechanics are responsible for installing and maintaining mechanical insulation on different types of plumbing, equipment, and HVAC systems. In order to perform work effectively on a plumbing system, an insulation mechanic must be able to identify the type of plumbing system being serviced. Recognizing common types of plumbing systems, understanding which systems do or do not require mechanical insulation, and anticipating which types of flanges, valves, and other fittings to expect in a system will greatly enhance a mechanical insulation worker's ability to efficiently plan and perform the work.

Module 19209

Trainees with successful module completions may be eligible for credentialing through the NCCER Registry. To learn more, go to **www.nccer.org** or contact us at 1.888.622.3720. Our website has information on the latest product releases and training, as well as online versions of our *Cornerstone* magazine and Pearson's product catalog.

Your feedback is welcome. You may email your comments to **curriculum@nccer.org**, send general comments and inquiries to **info@nccer.org**, or fill in the User Update form at the back of this module.

This information is general in nature and intended for training purposes only. Actual performance of activities described in this manual requires compliance with all applicable operating, service, maintenance, and safety procedures under the direction of qualified personnel. References in this manual to patented or proprietary devices do not constitute a recommendation of their use.

19209 V2

Objective

When you have completed this module, you will be able to do the following:

1. Identify basic types of plumbing systems.
   a. Identify common types of cold water plumbing systems.
   b. Identify common types of hot water plumbing systems.
   c. Identify various types of plumbing systems such as drainage, reclaimed water, and plenum systems.

Performance Tasks

This is a knowledge-based module; there are no Performance Tasks.

Trade Terms

Backflow preventer
Booster pump
Branch
Chase
Distribution system
Main
Non-potable water
Plenum

Potable water
Recirculating hot water
Refrigeration
Riser
Runout
Trapeze hangers
Water meter
Water softener

Industry Recognized Credentials

If you are training through an NCCER-accredited sponsor, you may be eligible for credentials from NCCER's Registry. The ID number for this module is 19209. Note that this module may have been used in other NCCER curricula and may apply to other level completions. Contact NCCER's Registry at 888.622.3720 or go to **www.nccer.org** for more information.

Contents

Figures

1.0.0 PLUMBING SYSTEM IDENTIFICATION

Objective

Identify basic types of plumbing systems.

a. Identify common types of cold water plumbing systems.
b. Identify common types of hot water plumbing systems.
c. Identify various types of plumbing systems such as drainage, reclaimed water, and plenum systems.

Trade Terms

Backflow preventer: A device that prevents liquid from flowing backward in the event of a loss in pressure.

Booster pump: A pump that increases the pressure of a liquid to provide adequate pressure for the upper floors of a facility.

Branch: A part of a distribution piping system that is usually smaller than a main and is used to connect a main to two or more runouts.

Chase: A vertical enclosure that houses piping and ductwork in a structure.

Distribution system: A piping system consisting of mains, risers, drops, branches, tanks, pumps, valves, fixture connections, and additional equipment.

Main: The principal distribution piping that connects a supply source to all of the branches in a system.

Non-potable water: Water that is not suitable for drinking, such as graywater / reclaimed water.

Plenum: An enclosed space in a facility, often located above a workspace between the structural ceiling and a drop-down ceiling, which allows air circulation for HVAC systems.

Potable water: Water that is suitable for drinking (for example, domestic water, city water, or well water).

Recirculating hot water: A system in which hot water is pumped in a continuous cycle to provide instant hot water at a fixture.

Refrigeration: The process of transferring heat from one substance or area to another in order to lower the temperature of the substance or area.

Riser: Vertical piping to or from a main, branch, or runout.

Runout: Piping to or from a branch or main to a plumbing unit or fixture connection, a heating and/or cooling unit connection, or a process equipment connection.

Trapeze hangers: Pipe hangers with parallel vertical rods that are suspended from a structure and connected at the bottom with a horizontal member from which one or more pipes can be supported.

Water meter: A device that measures and records the amount of water that passes through a pipe or similar object.

Water softeners: Pieces of equipment that remove minerals from water.

Plumbing pipe systems in commercial facilities that mechanical insulators typically work on include domestic cold water, domestic hot water, recirculating hot water, sanitary drainage, storm drainage, vents, condensate drains, and, in some cases, reclaimed water.
Not every piping system in a facility requires mechanical insulation. For instance, systems that operate at ambient temperature (i.e., the temperature of the surrounding air) do not require mechanical insulation since there is no loss of heat energy or condensation formation. These include gas and vacuum piping systems. However, other pipe systems and equipment may require insulation, such as a hot water generator or converter, water storage tank, booster pump, circulating pump, or water softener. This module examines several common pipe systems, including cold water piping, hot water piping, and some other pipe systems that insulation mechanics are likely to encounter.

> **WARNING!**
>
> Always be familiar with the type of insulation you are working with. Asbestos is a hazardous substance once commonly used as an insulating material. The use of asbestos as an insulating material was banned in the United States in 1975, but insulation workers may still encounter asbestos insulation in older facilities. Any insulation worker who is expected to handle and remove asbestos must complete a training program that is accredited by the US Environmental Protection Agency and wear special proper personal protective equipment (PPE).

1.1.0 Cold Water Piping

Plumbing cold water piping includes domestic water, potable water, city water, and well water. Whatever the designation, cold water piping that operates below ambient temperature typically requires mechanical insulation to prevent condensation from forming on the pipes. Some examples of equipment that cold water piping connects include the following:

- Bathroom fixtures (sinks, toilets, showers)
- Kitchen items (sinks, dishwashers, bar fixtures, icemakers)
- Janitor sinks
- Hose bibs
- Laboratory tables
- Can washing equipment
- Eye-wash stations
- Emergency showers

1.1.1 Cold Water Piping Materials

The materials used to make cold water piping have evolved over the years. Some of the most common materials used in the past include cast iron (usually underground and up to the point of entry into a facility), galvanized steel, copper, and various types of plastic.

Over time, some piping materials developed problems. For instance, galvanized steel pipes allow the buildup of sediment on the inside of the pipe. This gradually reduces the inside diameter of the pipe. Eventually, the buildup becomes so severe that water can barely flow through the pipe. Galvanized pipe also has a tendency to corrode at connecting joints, leading to leaks. For these reasons, galvanized steel has been largely replaced by more reliable materials.

Cast iron piping is a long-lasting material that is still in use, more so for drain piping than supply piping. While cast iron piping can last for decades, it is susceptible to corrosion and rust, breaks from contact by equipment, and splits from freezing. *Figure 1* shows a badly corroded cast iron water pipe.

Traditionally, the most widely used plumbing pipe has been copper tubing (*Figure 2*). Copper resists corrosion, lasts for a long time, is available in many sizes and lengths, and is relatively easy to work with.

While copper is a popular pipe material, it has some disadvantages. For instance, copper connections may need to be soldered or brazed. In addition, copper is expensive. Other piping materials can often be purchased at a fraction of the cost of copper. For this reason, many modern cold water systems use rigid plastic pipe made from polyvinyl chloride (PVC) and chlorinated polyvinyl chloride (CPVC), as well as flexible tubing made from cross-linked polyethylene (PEX). It is also common to use different types of piping in the same system.

Figure 1 Corroded cast iron pipe.

Figure 2 Copper tubing installation in office facility.

1.1.2 Cold Water System Layout

Figure 3 shows the basic layout and components of a typical cold water piping system as it enters a facility from an underground city water line. The piping used at the entry area is often the same material as the underground service pipe, which is frequently cast iron. Once inside the facility, the piping changes to the type of material used throughout the internal piping system, such as copper or plastic.

A main shut-off valve provides a way to isolate the facility water system from the city water supply. Immediately after the main shut-off valve is another valve called a backflow preventer. It is used to prevent water from the facility system from flowing back into the city water system if the city

water system were to lose pressure. A water meter is used to measure the amount of water used by the facility. Water meters are sometimes located outside of the facility in a meter pit or manhole.

In a multi-story facility, the piping may connect to booster pumps (*Figure 4*) to provide adequate water pressure for the upper floors. The system usually includes valves, strainers, and flexible connections to reduce the pump vibration transmitted to the piping system. Some cold water piping systems might also include a water softener. Water softeners are usually factory-assembled units that may be factory insulated. The softeners, booster pumps, and water meters are usually located in an area near where the city water main enters the facility.

In addition to providing domestic cold water, cold water piping systems are commonly used to provide fill water or makeup water for other systems. Some examples of these other systems include the following:

- Boiler systems
- Chilled water pipe systems
- Distilled water systems
- Heating pipe systems
- Condenser water systems
- Refrigerated drinking water systems
- Ice machines
- Non-potable water systems

In most facilities, the cold water piping and hot water piping run side-by-side throughout the facility. The hot pipe might be smaller in diameter than the cold pipe. Because water will flow faster

Recycling Piping Materials

GOING GREEN

Many piping materials are recyclable. For instance, copper, PVC, and CPVC can be recycled to make new piping or other products. While PEX tubing is not recyclable, it can be reprocessed by grinding it into a powder and using it as an additive in other plastics. Recycling and reprocessing piping materials reduces landfill waste, saves money, and helps the environment.

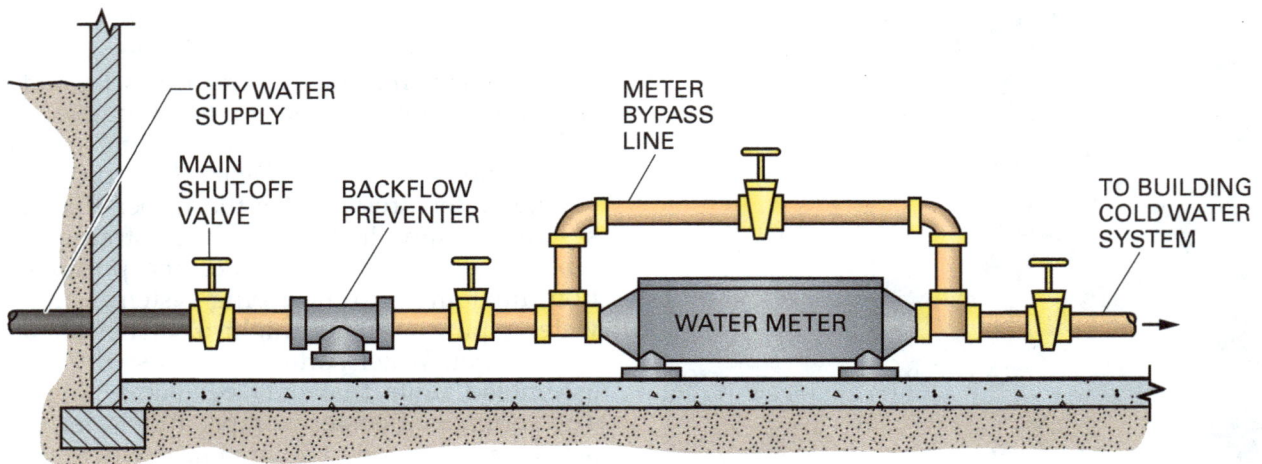

Figure 3 Cold water system at entry to facility.

through a smaller pipe, hot water will be delivered to the fixture more quickly, resulting in less wasted water and energy.

By convention, when facing a fixture, the cold pipe is on the right and the hot is on the left. *Figure 5* is a simplified illustration showing the various parts of the distribution system.

The main piping components of the distribution system include the following:

- *Main* – The principal distribution piping that connects a supply source to every branch in a system.
- *Branch* – A part of a distribution piping system that is usually smaller than a main and used to connect a main to two or more runouts. (The final piping connection between a runout and a plumbing fixture is NOT a branch.)

Figure 4 Booster pump.

- *Riser* – Vertical piping to or from a main, branch, or runout.
- *Runout* – Piping to or from a branch or main to a plumbing unit or fixture connection, a heating and/or cooling unit connection, or a process equipment connection.

Generally, plumbing piping is installed in concealed locations. For example, mains and branches might be located in a facility's basement or above ceilings. Risers are commonly located behind walls or in a vertical enclosure called a chase. Chases are situated behind areas where several plumbing fixtures are located, such as toilets in a public restroom. There are also fixture connections located in walls or chases behind fixtures.

Since plumbing pipes must be insulated after they have been tested but before any plumbing fixtures are installed, it is important to identify the various systems to determine the mechanical insulation requirements. As stated earlier, a lot of plumbing water pipe is copper. However, plastic pipe has become more popular since most building codes have been revised to allow its use.

1.1.3 Pipe System Components

If several pipe systems are installed side-by-side, you may encounter trapeze hangers (*Figure 6*) supporting all of the pipes. A single trapeze hanger can eliminate the need for several individual hangers.

Whatever hangers are used, the number of hangers required will depend on the rigidity of the pipe. Smaller pipes and plastic pipes are less rigid, so when working with those pipes you will encounter more hangers than with larger pipes and metal pipes.

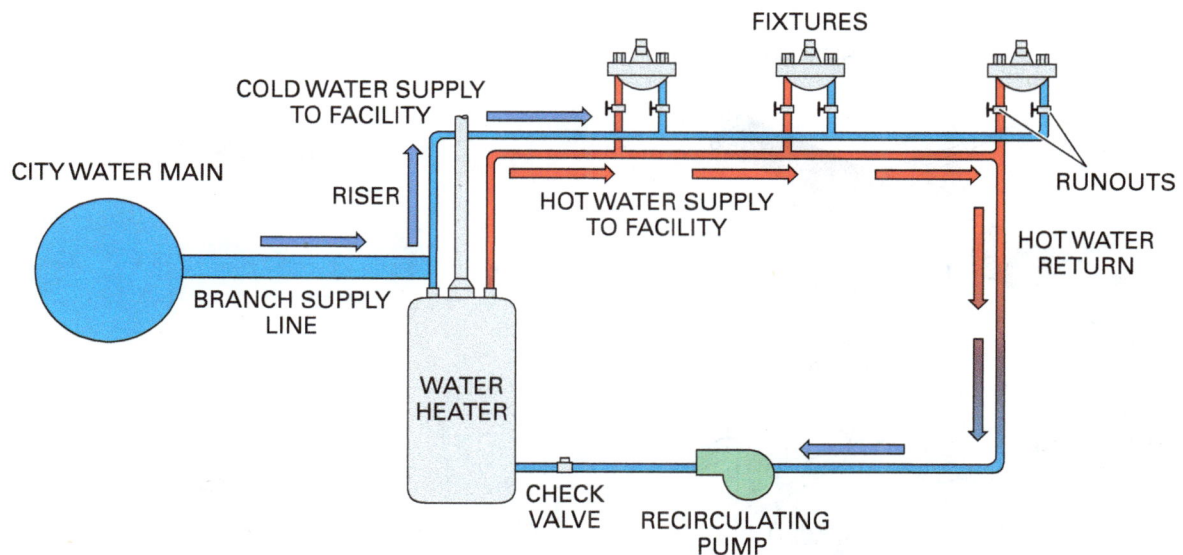

Figure 5 Plumbing water distribution system.

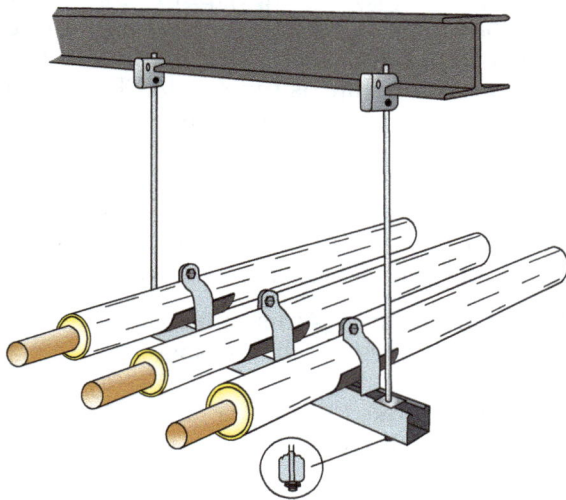

Figure 6 Trapeze hanger.

Where plumbing pipe passes through walls and floors, a pipe sleeve or penetration seal known as *firestopping* (*Figure 7*) is used to create a seal between the piping and the wall passage. Whatever device is used should be large enough to permit the pipe with its mechanical insulation and vapor retarder to pass through. If the walls and floors are fire rated, the area where the pipe passes through the wall will normally be filled and sealed with a firestopping material. This is done to maintain the fire rating of the wall that is penetrated.

Plumbing cold water systems may include other devices that a mechanical insulator must be able to recognize. For instance, water hammer arrestors (*Figure 8*) may be located at the end of main runs and at fixtures to reduce noise or shock that occurs when the water flow is suddenly stopped by closing of a valve or faucet.

Not only is a water hammer noisy, it can be a very destructive force. A water hammer, or hydraulic shock wave, can break pipes, instruments, valves, and other piping system components.

Insulators are also likely to see dielectric connectors (*Figure 9*), also called *dielectric unions*, in systems where pipes made from different materials are joined together. These connectors are used to prevent electrical currents caused by dissimilar metals touching in the presence of water. If pipes of dissimilar metals are joined without a dielectric fitting, the electrical currents will quickly cause the joint to fail.

> **NOTE**
>
> Do not disturb any telephone ground cables attached to the cold water main. Install mechanical insulation directly over the ground(s).

Exposed plumbing water connections at fixtures are typically made with a final, aesthetic finish such as chrome, stainless steel, or nickel. These types of piping are used for the sake of appearance and usually do not require mechanical insulation. There is, however, an exception that relates to the 1990 Americans with Disabilities Act (ADA). ADA regulations state that "Hot water and drain pipes exposed under sinks shall be insulated or otherwise configured so as to protect against contact. There shall be no sharp or abrasive surfaces under sinks." To provide a way to comply with the ADA regulation, some companies make special pipe covers that can be attached to under sink piping at handicap accessible sinks to shield users from hot surfaces and sharp edges.

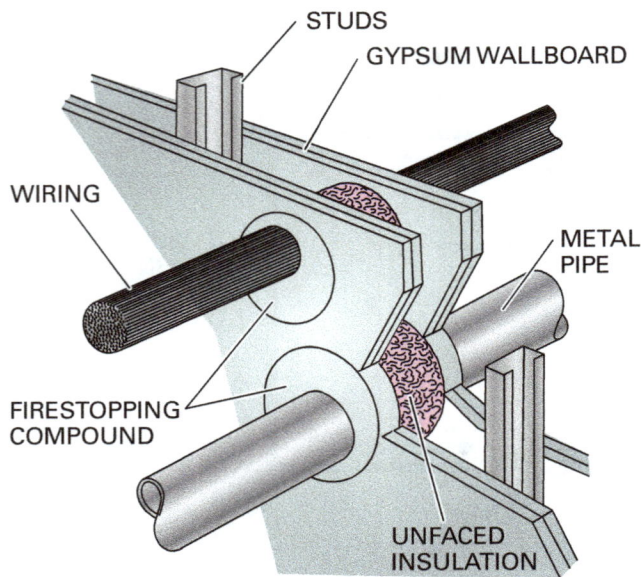

WALLBOARD PENETRATION

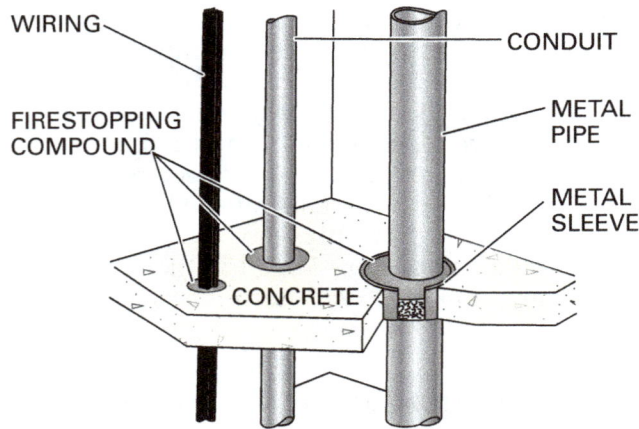

CONCRETE PENETRATION

Figure 7 Mechanical fire stop device.

Figure 8 Water hammer arrestors.

Figure 9 Dielectric connector.

1.2.0 Hot Water Piping

Plumbing hot water is simply plumbing cold water that has been heated, typically to a temperature between 120°F and 150°F (49°C and 66°C). Some systems operate as high as 180°F (82°C). Because the hot water is above ambient temperature, hot water piping should be insulated to conserve energy, reduce the temperature drop throughout the system, and protect personnel from hot pipes. Some examples of equipment that plumbing hot water piping connects include the following:

- Bathroom fixtures (sinks, showers)
- Kitchen items (sinks, dishwashers)
- Janitor sinks
- Laboratory sinks
- Car wash equipment

> **NOTE**
>
> Plumbing hot water is not used for hot water heating systems. These systems provide their own heat source.

1.2.1 Hot Water Piping Materials

Plumbing hot water systems use the same types of pipe material as plumbing cold water systems: copper, PVC, CPVC, and PEX. As with cold water piping, visible fixture connections are typically made with a decorative finish such as chrome, stainless steel, or nickel. In some applications, such as the PEX tubing connected to the manifold in *Figure 10*, the piping is color coded—red for hot and blue for cold.

Figure 10 Color-coded hot and cold PEX tubing leaving manifold.

1.2.2 *Hot Water System Layout*

The equipment used to convert plumbing cold water into plumbing hot water may be similar to a residential hot water heater in that it both heats and stores the water, and is factory-insulated and jacketed. A facility may be served from a central hot water system or by local systems in several areas. Either way, one or more domestic hot water heaters are needed. The size and location of domestic hot water heaters will depend on the facility's needs and economics.

Commercial water heaters may also serve as storage tanks. In large commercial systems, a separate vessel might be used as the storage tank. In such systems, the plumbing cold water connection is at the bottom of the tank while the hot water main is connected to the top of the storage tank. As the water is heated, it rises so that the hottest water is located at the top of the tank. Whether a water heater is used as the storage tank or a separate storage tank is used, mechanical insulation should be used to prevent heat loss.

Water heaters may be fueled by electricity, gas, oil, or solar power, or they may be augmented by a separate hot water heating system. The heaters may be self-contained package units with factory

Solar Power

GOING GREEN

The number of commercial facilities that use solar energy for heating water is growing. Solar is a renewable energy form that does not cause pollution, such as greenhouse gases. Although startup costs can be high, the ongoing cost of operation is minimal and the systems last for many years. In a common solar water heating arrangement, water from a city's system flows to a large storage tank. Inside the tank is a heat exchanger coil that is filled with a heat transfer fluid such as a propylene glycol mixture. A pump is used to circulate the heat transfer fluid through a solar collector where the fluid absorbs heat. The heated fluid is then pumped through the coils in the water storage tank, where it transfers heat to the water. This is a closed-loop system. The heated water from the storage tank is then used by the facility. Some systems are capable of heating water to 170°F (77°C) or more.

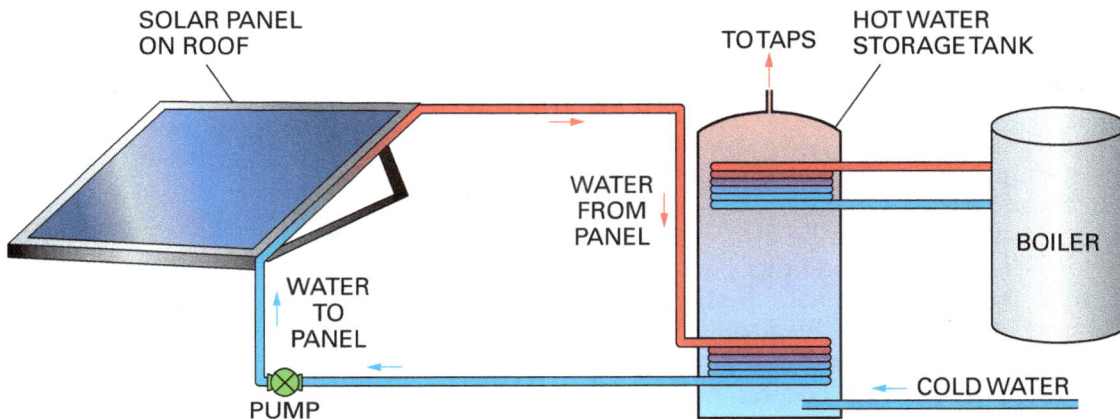

installed mechanical insulation, or they may require field-installed mechanical insulation.

Many hot water systems eliminate the need for storage tanks by using tankless heaters. These heaters create hot water instantly by using ceramic electric heaters or by burning gas. These systems provide hot water to fixtures in less time than conventional hot water heaters, which saves water.

Plumbing hot water piping starts from the hot water heating unit and runs throughout the facility in exactly the same manner as the plumbing cold water piping. Insulation being installed on mains, branches, risers, runouts (fixture connections), and in any concealed area is installed the same way as on cold water systems. The two systems commonly run side-by-side, although the diameter of the hot water pipe might be smaller than that of the cold water pipe. By convention, at plumbing fixtures (sinks, tubs, or showers) the hot pipe is on the left.

Pipe hangers, supports, anchors, and sleeves are basically identical for both hot and cold water piping. Although the integrity of the vapor retarder on cold pipe mechanical insulation is important, it is not crucial on hot water pipe. However, while expansion is negligible and of no consequence with cold water pipe, hot water pipe must provide for thermal expansion. Insulation mechanics are likely to encounter two common methods for dealing with thermal expansion: the use of an expansion loop and the use of an expansion tank.

Plumbing hot water systems in commercial applications generally include recirculating or return piping. Recirculating piping allows the hot water to be available at any outlet as soon as it is turned on. Without a recirculating pipe, the entire distribution system to that particular fixture would have to fill up with hot water before it comes to the fixture. To keep the hot water moving in the system, a recirculating pump is connected near the heater. *Figure 11* shows a simplified diagram of recirculating hot water piping. The recirculating or return piping should be insulated the same as the plumbing hot water system.

Mechanical insulators should be aware that facilities in which dishwashers and other equipment is present may require water at a higher temperature than what is furnished by a conventional plumbing hot water system. The higher temperatures are achieved by using a booster heater located near the equipment that requires the higher-temperature water. The higher-temperature water (180°F [82°C] or higher) is supplied by a separate piping system through the booster heater.

1.3.0 Piping Systems for Drainage, Reclaimed Water, and Plenums

Mechanical insulation installers not only work with cold water and hot water systems, but also with numerous other piping systems. Some of the more common systems include storm drain piping, condensate drain piping, piping used in reclaimed water systems, and piping in a return air plenum.

1.3.1 Storm Drain Piping

Storm drain piping is often referred to as *roof drains*, *downspouts*, *roof leaders*, or *area drains*. Because there may be confusion as to the intent and the scope of the piping to be insulated, care must be taken when interpreting the mechanical insulation specification.

Generally, a storm drain system carries water from a roof to an underground water system or a reclaim tank. Reclaimed water is treated to make non-potable water that can be recycled for irrigation, toilet flushing, process cooling, and similar uses. (More detail about reclaimed water piping is provided later in this section.) The drains, downspouts, gutters, or leaders are actually parts of the storm drainage system. Mechanical insulation may be required where the piping runs through an indoor space or where it runs above equipment because whenever cold drainage water runs through the pipes, it could cause condensation to form on the piping. *Figure 12* shows a typical storm drain system.

Storm drain piping may be any of the following types:

- Plastic pipe
- Cast iron pipe
- Welded steel pipe

If cast iron pipe is used, the joints and fittings may be of the bell-and-spigot type or the rubber gasket stainless steel banded type. Hangers should be similar to the hangers used on cold water and hot water piping and permit the mechanical insulation on the pipe to pass through the hangers continuously. Due to the large pipe size and weight of drainage systems, contact hangers are used. This requires close attention to the mechanical insulation and vapor seal at the hanger points.

Storm drainage systems may have cleanout type fittings and long radius sweeps or ells. The sump, which is the bottom part of the roof drain, and the drain supports usually have irregular configurations. If the drain piping requires mechanical insulation, the sump should be included.

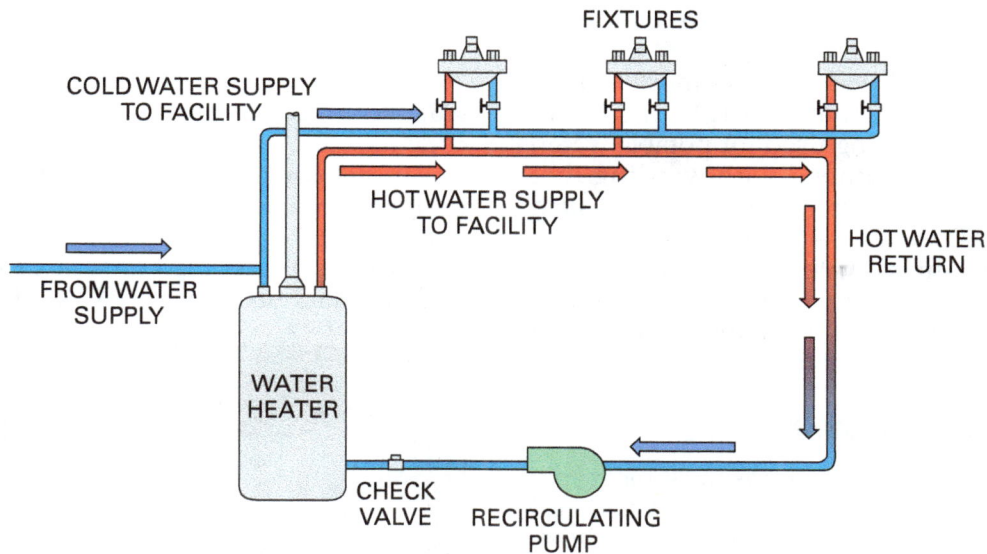

Figure 11 Plumbing system with recirculating hot water.

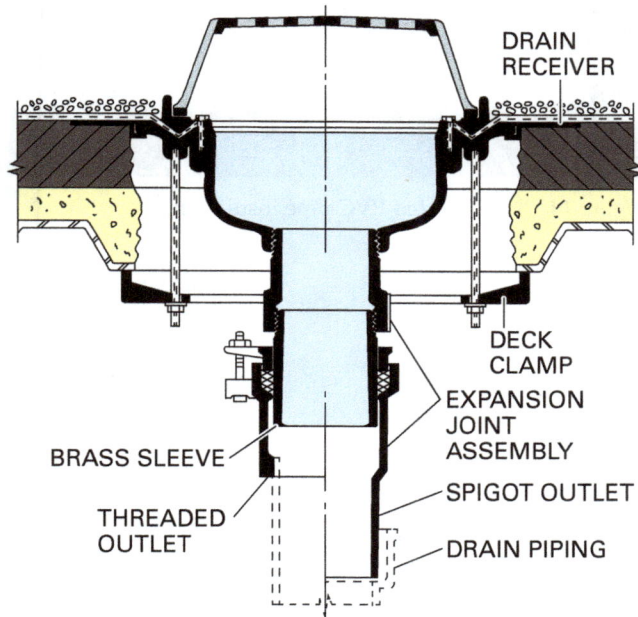

Figure 12 Storm drain system.

The storm system may have roof drains with connecting piping that terminates in the underground system. Roof drainage may be designed to spill over the side of the roof into gutters, which may or may not drain into an underground system. There may be variations of both systems, and the mechanical insulation requirements will vary accordingly.

1.3.2 Condensate Drain Piping

Condensate drain piping is a piping system that is connected to air conditioning units, refrigeration units, and any cooling coil. Often, the piping is connected to drain pans underneath the equipment that collect the condensate that forms on the coils when ambient air passes through the cooling coil. The condensate is approximately the same temperature as the coil and the drain piping requires mechanical insulation to control the condensation that may occur on the cold pipe. Generally, the condensate is piped to a floor drain, a dry-well, or a splash-block outside the facility.

Condensate drain system pipe is usually either copper tubing or plastic piping. A P-trap is often used to prevent gas and odor from passing through the floor drain or waste system into the facility. Other than P-traps, the only fittings used are standard elbows and tees. There are no valves in condensate drain systems. Condensate drain piping is installed by plumbing, air conditioning, or mechanical contractors.

1.3.3 Reclaimed Water Piping

Reclaimed water is one of several terms used to describe wastewater that is collected from various sources and recycled to meet certain non-potable water needs. The wastewater that is collected is commonly referred to as *graywater*. In commercial and industrial applications, graywater can come from sources such as sinks, showers, different types of washing machines, as well as roof and area drains. Graywater does not, however, come from toilets. Since graywater is relatively less contaminated than sewage wastewater, it can be filtered and treated with less aggressive methods. The result is non-potable water with enough purity to be reused for irrigation, toilet flushing, some heat transfer processes, and replenishing groundwater. The use of graywater can save companies and municipalities millions of gallons of water each year.

Piping for reclaimed water is commonly made of PVC, CPVC, or PEX that is purple in color to distinguish it from piping used for potable water. The piping is often stamped with wording that alerts people that the water is non-potable, such as "RECLAIMED WATER—DO NOT DRINK". In most cases, reclaimed water piping is not insulated. If it is determined that insulation should be used, the same mechanical insulation that is used for hot water piping is suitable.

1.3.4 Piping in Return Air Plenums

Piping used in return air plenums must meet specific criteria based on *ASTM E 84, Standard Test Method for Surface Burning Characteristics of Building Materials.* This standard states that any material used in a plenum must be noncombustible or meet some stringent requirements that pertain to flame spreading and smoke development.

Copper tubing meets the ASTM requirement since it is a noncombustible material. Plastic piping is different. Building codes relating to plastic piping in return air plenums vary. But as a general rule, exposed PVC, CPVC, and PEX piping do not meet the flame spreading and smoke development requirements of the ASTM standard. These piping materials must be covered with a suitable insulating material that meets the fire code requirements.

Various companies produce insulation that is designed for use in plenums. Most of these products use generic descriptions such as plenum wrap or fire barrier plenum wrap. *Figure 13* shows plenum-rated insulation for PVC pipe.

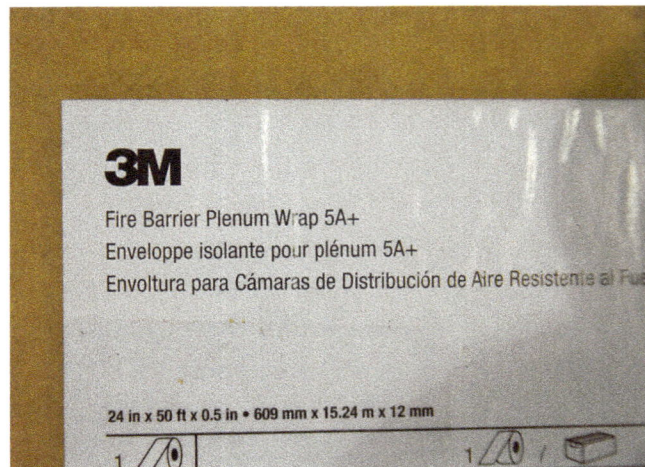

Figure 13 Plenum-rated PVC pipe insulation.

1.0.0 Section Review

1. Which of the following is a disadvantage of using copper pipe?

    a. Copper is soft and requires special handling.
    b. Copper is subject to corrosion with a special coating.
    c. Copper may need to be soldered or brazed.
    d. Copper will cause other metals around it to corrode.

2. Hot water piping operates above ambient temperatures, so it requires mechanical insulation to _____.

    a. conserve energy by reducing heat loss
    b. cool the piping to prevent drooping
    c. absorb any moisture from condensation
    d. differentiate it from cold water piping

3. A piping system that is often connected to drain pans underneath air conditioning and refrigeration units is a(n) _____.

    a. recirculating hot water system
    b. condensate drain system
    c. reheat makeup water system
    d. ambient water system

SUMMARY

The ability to recognize and understand plumbing systems is a vital part of a mechanical insulator's job. Plumbing systems can vary a great deal. Recognizing common types of plumbing systems, understanding which systems do or do not require mechanical insulation, and anticipating which types of components to expect in a system will greatly enhance a mechanical insulation worker's ability to plan and perform the work.

Once you can recognize the components of a plumbing system you can anticipate the types of insulation that will be required and the sequence of insulation. This knowledge will help you plan your work for maximum efficiency.

1. Which of these is a reason why systems that operate at ambient temperature are not insulated?
   a. They have no source of moisture.
   b. They do not experience heat loss.
   c. Additional pressure will loosen the insulation.
   d. Air at ambient temperature causes a loose fit.

2. The word potable is used to describe _____.
   a. any solid material that is edible
   b. semi-pure water that can be used for flushing toilets
   c. a liquid that is suitable for drinking
   d. untreated water in a sewage treatment facility

3. Traditionally, the most widely used pipe in plumbing systems is _____.
   a. PVC pipe
   b. copper tubing
   c. PEX tubing
   d. steel pipe

4. To ensure adequate water pressure for the upper floors of a multi-story facility, water piping may be connected to _____.
   a. backflow preventers
   b. manifold valves
   c. vacuum tanks
   d. booster pumps

5. A vertical enclosure that houses piping and ductwork and is usually situated behind an area where several plumbing fixtures are located is known as a _____.
   a. chase
   b. runout
   c. riser
   d. manifold

6. When several piping systems are installed side-by-side, multiple pipes can be supported using a _____.
   a. clevis hanger
   b. contact hanger
   c. strap hanger
   d. trapeze hanger

7. Which of these statements about plumbing hot water is true?
   a. It is commonly used to heat commercial facilities.
   b. It is simply plumbing cold water that has been heated.
   c. It must be heated above 212°F (100°C) to be considered plumbing hot water.
   d. Its piping is insulated only to protect personnel from hot surfaces.

8. Compared to plumbing cold water systems, the pipe hangers and supports for plumbing hot water systems _____.
   a. must be farther apart
   b. use different materials
   c. should have heat tracing
   d. are basically identical

9. Storm drain piping is usually insulated because cold drainage water could cause _____.
   a. condensation to form on the piping
   b. evaporation to occur in adjacent vessels
   c. a chemical reaction on the surface of the pipes
   d. ambient temperatures to drop below specifications

10. If PVC pipe is used in an air return plenum, it must be _____.
    a. white in color and labeled HOT
    b. secured to the bottom of the plenum
    c. covered with a suitable insulating material
    d. kept bare to allow visible identification

Trade Terms Quiz

Fill in the blank with the correct term that you learned from your study of this module.

1. _____ can be used to support several pipes.

2. A(n) _____ removes minerals from water.

3. A(n) _____ system pumps hot water in a continuous cycle to provide instant hot water at a fixture.

4. Water that is suitable for drinking is called _____.

5. The principal distribution piping that connects a supply source to all system branches is a(n) _____.

6. A(n) _____ prevents material from flowing backward.

7. A vertical enclosure for piping and ductwork is called a(n) _____.

8. _____ transfers heat from one object to another or one area to another.

9. A(n) _____ is frequently needed to provide adequate pressure to ensure water flow in the upper floors of a facility.

10. An enclosed space between the structural ceiling and a drop-down ceiling in a facility that allows air circulation for HVAC systems is a(n) _____.

11. A(n) _____ measures the flow of water through a pipe.

12. A(n) _____ connects a main to two or more runouts.

13. A facility's _____ includes all the piping in the facility.

14. The piping that connects a branch or main to the final use equipment is a(n) _____.

15. A vertical pipe in a piping system is referred to as a(n) _____.

16. Graywater and reclaimed water are forms of _____.

Trade Terms

Backflow preventer	Distribution system	Potable water	Runout
Booster pump	Main	Recirculating hot water	Trapeze hangers
Branch	Non-potable water	Refrigeration	Water meter
Chase	Plenum	Riser	Water softener

Trade Terms Introduced In This Module

Backflow preventer: A device that prevents liquid from flowing backward in the event of a loss in pressure.

Booster pump: A pump that increases the pressure of a liquid to provide adequate pressure for the upper floors of a facility.

Branch: A part of a distribution piping system that is usually smaller than a main and is used to connect a main to two or more runouts.

Chase: A vertical enclosure that houses piping and ductwork in a structure.

Distribution system: A piping system consisting of mains, risers, drops, branches, tanks, pumps, valves, fixture connections, and additional equipment.

Main: The principal distribution piping that connects a supply source to all of the branches in a system.

Non-potable water: Water that is not suitable for drinking, such as graywater / reclaimed water.

Plenum: An enclosed space in a facility, often located above a workspace between the structural ceiling and a drop-down ceiling, which allows air circulation for HVAC systems.

Potable water: Water that is suitable for drinking (for example, domestic water, city water, or well water).

Recirculating hot water: A system in which hot water is pumped in a continuous cycle to provide instant hot water at a fixture.

Refrigeration: The process of transferring heat from one substance or area to another in order to lower the temperature of the substance or area.

Riser: Vertical piping to or from a main, branch, or runout.

Runout: Piping to or from a branch or main to a plumbing unit or fixture connection, a heating and/or cooling unit connection, or a process equipment connection.

Trapeze hangers: Pipe hangers with parallel vertical rods that are suspended from a structure and connected at the bottom with a horizontal member from which one or more pipes can be supported.

Water meter: A device that measures and records the amount of water that passes through a pipe or similar object.

Water softeners: Pieces of equipment that remove minerals from water.

Additional Resources

This module presents thorough resources for task training. The following reference material is recommended for further study.

The Mechanical Insulation Best Practices Guide, Thermal Insulation Association of Canada. Available at **www.tiac.ca/en/resources/best-practices-guide**

Mechanical Insulation Design Guide, National Institute of Building Sciences. Available at **www.wbdg.org/guides-specifications/mechanical-insulation-design-guide**

The following websites offer resources for products and training:

    Copper Development Association, Inc., **www.copper.org**

    Tradesman Supply, LLC. "PEX Information." **www.pexinfo.com**

    Piping Technology & Products, Inc., **www.pipingtech.com**

Figure Credits

Section Review Answer Key

Answer Section One	Section Reference	Objective
1. c	1.1.1	1a
2. a	1.2.0	1b
3. b	1.3.2	1c

NCCER CURRICULA — USER UPDATE

NCCER makes every effort to keep its textbooks up-to-date and free of technical errors. We appreciate your help in this process. If you find an error, a typographical mistake, or an inaccuracy in NCCER's curricula, please fill out this form (or a photocopy), or complete the online form at **www.nccer.org/olf**. Be sure to include the exact module ID number, page number, a detailed description, and your recommended correction. Your input will be brought to the attention of the Authoring Team. Thank you for your assistance.

Instructors – If you have an idea for improving this textbook, or have found that additional materials were necessary to teach this module effectively, please let us know so that we may present your suggestions to the Authoring Team.

NCCER Product Development and Revision

13614 Progress Blvd., Alachua, FL 32615

Email: curriculum@nccer.org
Online: www.nccer.org/olf

❏ Trainee Guide   ❏ Lesson Plans   ❏ Exam   ❏ PowerPoints   Other _____

Craft / Level: _____ Copyright Date: _____

Module ID Number / Title: _____

Section Number(s): _____

Description: _____

_____

_____

_____

_____

Recommended Correction: _____

_____

_____

_____

_____

Your Name: _____

Address: _____

_____

Email: _____ Phone: _____

This page is intentionally left blank.

Chilled and Hot Water Heating Systems

OVERVIEW

The bulk of the mechanical insulation work in any building is usually concentrated in the heating and air conditioning systems. This module introduces the components that go together to heat or cool a facility. Regardless of the design, these components require a great deal of insulation. The craftsperson must understand how the systems are designed in order to do a complete insulation job.

Module 19210

Trainees with successful module completions may be eligible for credentialing through the NCCER Registry. To learn more, go to **www.nccer.org** or contact us at 1.888.622.3720. Our website has information on the latest product releases and training, as well as online versions of our *Cornerstone* magazine and Pearson's product catalog.

Your feedback is welcome. You may email your comments to **curriculum@nccer.org**, send general comments and inquiries to **info@nccer.org**, or fill in the User Update form at the back of this module.

This information is general in nature and intended for training purposes only. Actual performance of activities described in this manual requires compliance with all applicable operating, service, maintenance, and safety procedures under the direction of qualified personnel. References in this manual to patented or proprietary devices do not constitute a recommendation of their use.

19210 V2

CHILLED AND HOT WATER HEATING SYSTEMS

Objectives

When you have completed this module, you will be able to do the following:

1. Identify the elements of chilled and hot water systems.
   a. Identify the elements of chilled water systems.
   b. Identify the elements of hot water heating systems.

Performance Tasks

This is a knowledge-based module; there are no Performance Tasks.

Trade Terms

Chilled water   Conditioned air
Condensation   Hot water heating

Industry Recognized Credentials

If you are training through an NCCER-accredited sponsor, you may be eligible for credentials from NCCER's Registry. The ID number for this module is 19210. Note that this module may have been used in other NCCER curricula and may apply to other level completions. Contact NCCER's Registry at 888.622.3720 or go to **www.nccer.org** for more information.

Contents

Figures

1.0.0 CHILLED WATER COOLING AND HOT WATER HEATING SYSTEMS

Objective

Identify the elements of chilled and hot water systems.

a. Identify the elements of chilled water systems.
b. Identify the elements of hot water heating systems.

Trade Terms

Chilled water: Water at below-ambient temperature used for cooling (particularly in air conditioning systems or in processes). Typical chilled water temperature for comfort cooling applications is 42°F–55°F (7°C–13°C).

Condensation: The physical process by which a liquid is removed from its vapor by cooling (e.g., water vapor turns into a liquid upon contact with a cold surface).

Conditioned air: Air that has been treated to control its temperature, humidity, and/or cleanliness to meet the requirements of a conditioned space.

Hot water heating: Water heated by natural or artificial means used for structural comfort heating or in processes. Typical hot water temperature for comfort heating applications is 170°F–190°F (77°C–88°C).

Mechanical systems in commercial buildings require a variety of insulation products. Although most insulation work is performed after mechanical systems are installed, a skilled insulation worker needs a basic understanding of how these systems are constructed to know and understand the insulation challenges they represent.

Mechanical systems consisting of heating, ventilating, and air conditioning equipment are designed by engineers. Engineers are also responsible for the sanitary and storm water control facilities of the project. A mechanical engineer, electrical engineer, civil or site engineer, and structural engineer form a design team that answers to the architect. This group will specify the location, type, and quantity of insulation needed for the heating and cooling systems.

Large commercial buildings often rely on chilled water cooling for both safety and cost efficiency. Water is inexpensive, noncorrosive, and nontoxic. There is also the added safety advantage of not having to circulate a chemical refrigerant such as ammonia throughout the building. Refrigerants are far more sensitive to pipe size, pipe length, vertical risers, and other common piping characteristics. The chilled water system is a closed piping circuit. As a result, chemical treatment to control contamination or corrosion is not a significant problem.

The same building may also rely on water to heat the facility. An advantage of a hot water heating system is that the water temperature can be varied to meet the heating requirements based on the outdoor weather conditions. Both chilled water cooling systems and hot water heating systems require properly installed insulation in order to prevent unwanted gain or loss of heat.

1.1.0 Chilled Water Systems

Chilled water, by definition, is water below the ambient or surrounding temperature. It is used as a cooling medium in commercial and industrial air conditioning systems (*Figure 1*). Typically, water is cooled in a chiller—a mechanical refrigeration unit designed to cool water circulating through the system. The chilled water is typically distributed by a piping system to an air handling unit, which houses water coils in an air duct system, to a fan coil unit, or to coils within an induction ventilation system. As air circulates across these coils, it is cooled and a great deal of the moisture it contains is removed through condensation.

Chilled water systems generally operate at temperatures that range from 42°F–55°F (7°C–13°C). Because these temperatures are usually below the ambient temperature, all piping and components in the system require insulation to prevent condensation.

1.1.1 Chilled Water System Components

The major components of a chilled water system are chillers (*Figure 2*), chilled water pumps, an expansion tank, air separators, air handlers, and a condensing unit. The chiller and pumps are usually located in a mechanical equipment room, possibly in the same room as the heating equipment.

In a complex of several buildings, the chiller and heat-generating equipment may be housed in a separate building or a central energy facility, with the cooling and heating water distribution

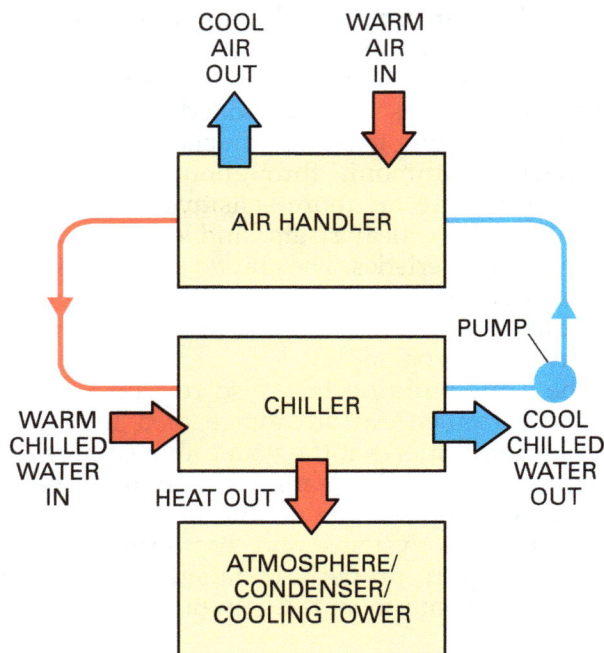

Figure 1 Basic schematic of a chilled water system.

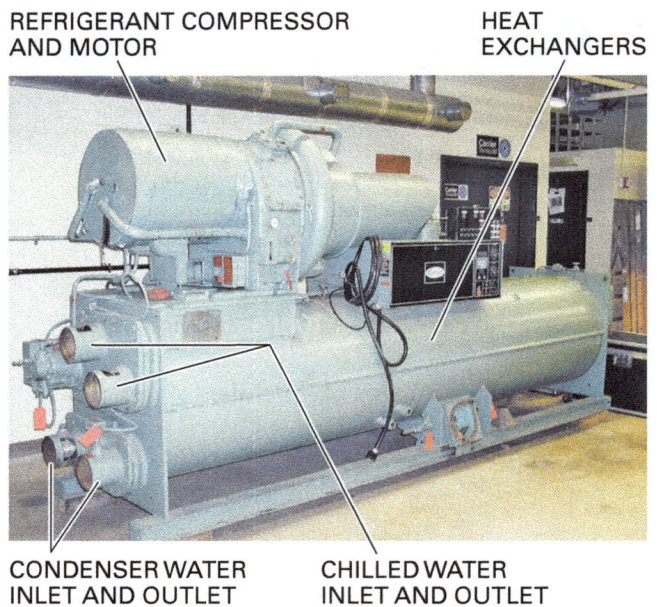

Figure 3 Water-cooled centrifugal chiller.

system located underground, above ground, or in tunnels. Air-cooled chillers, such as the one shown in *Figure 2*, are found outdoors. Water-cooled models (*Figure 3*) are typically located indoors.

Two or more chillers are used when an application is critical. If one chiller requires maintenance or repair, the system is never without some level of cooling. One popular approach today is the use of modular chillers, where a number of smaller chillers are connected to a common manifold.

Water-cooled chillers have one set of piping connections for the chilled water circulated for cooling purposes, and another set for condenser water. Chilled water supply and return lines are piped to the chilled water pumps (*Figure 4*) for distribution. Two pumps are often used to provide continuous operation in the case of a failure. Pumps are manufactured in many different configurations, depending on the size required and the space available. *Figure 4* provides an example of a centrifugal pump.

Chilled water systems are normally designed with an expansion tank, shown in *Figure 5* (*A*) that allows for changes in the system's water volume due to temperature changes, leaks, or added makeup water. A device called an *air separator*, shown in *Figure 5* (*B*), may be located near the chilled water pumps. The air separator removes air entrapped in the system which can lead to corrosion and pump impeller damage. The air separator may be as simple as a specialized fitting routed to the expansion tank, or it may be a large device such as the one shown in *Figure 5* (*A*), vented directly to the atmosphere. A chemical feed system may be connected near the chilled water pumps to provide water treatment for the chilled water.

Figure 6 shows a simple schematic of some of the basic components and piping connections in a chilled water plant. Note the many valves, strainers, and other devices. The insulation installer must consider all these devices, as each component in the piping system represents a different challenge to the insulator.

There are flanged connections and valves at the entrance of the chillers, and flanged connections, valves, strainers, and vibration isolators

Figure 2 Typical outside chiller unit.

CENTRIFUGAL   PUMP-MOTOR   MOTOR
PUMP   COUPLING
                     GUARD

Figure 4 Centrifugal chilled water pump.

(A) DIAPHRAGM EXPANSION TANK

(B) AIR SEPARATOR

Figure 5 Expansion tank and air separator.

at the pipe connections to the pumps (*Figure 7*). There are also bypasses, chemical feed connections, thermometer wells, and control valves in the system. These components are all subject to condensation, and therefore should be insulated and sealed.

1.1.2 Chilled Water System Heat Exchangers

Chilled water will run to air handling equipment, fan coil units, or induction units, depending on the type of system. Air handling equipment may be large built-up units with field-installed cooling coils. These air handlers may be located in a machine room with the chilled water equipment or in a separate area (*Figure 8*).

The coils are piped inside or outside of the unit casing with unions, flanges, control valves, and shutoff valves. In some applications, air handling units (*Figure 9*) are located on the roof. These units supply conditioned air to the building through an internal duct distribution system. Rooftop units are usually factory-assembled and rigged from the delivery truck directly to the roof. The only mechanical connections required are supply and return ducts, chilled and hot water piping, and condensate drains.

Many air handling units are located indoors in equipment rooms or above suspended ceilings. Air handlers may serve one room or a complete floor, depending on the design. They may have

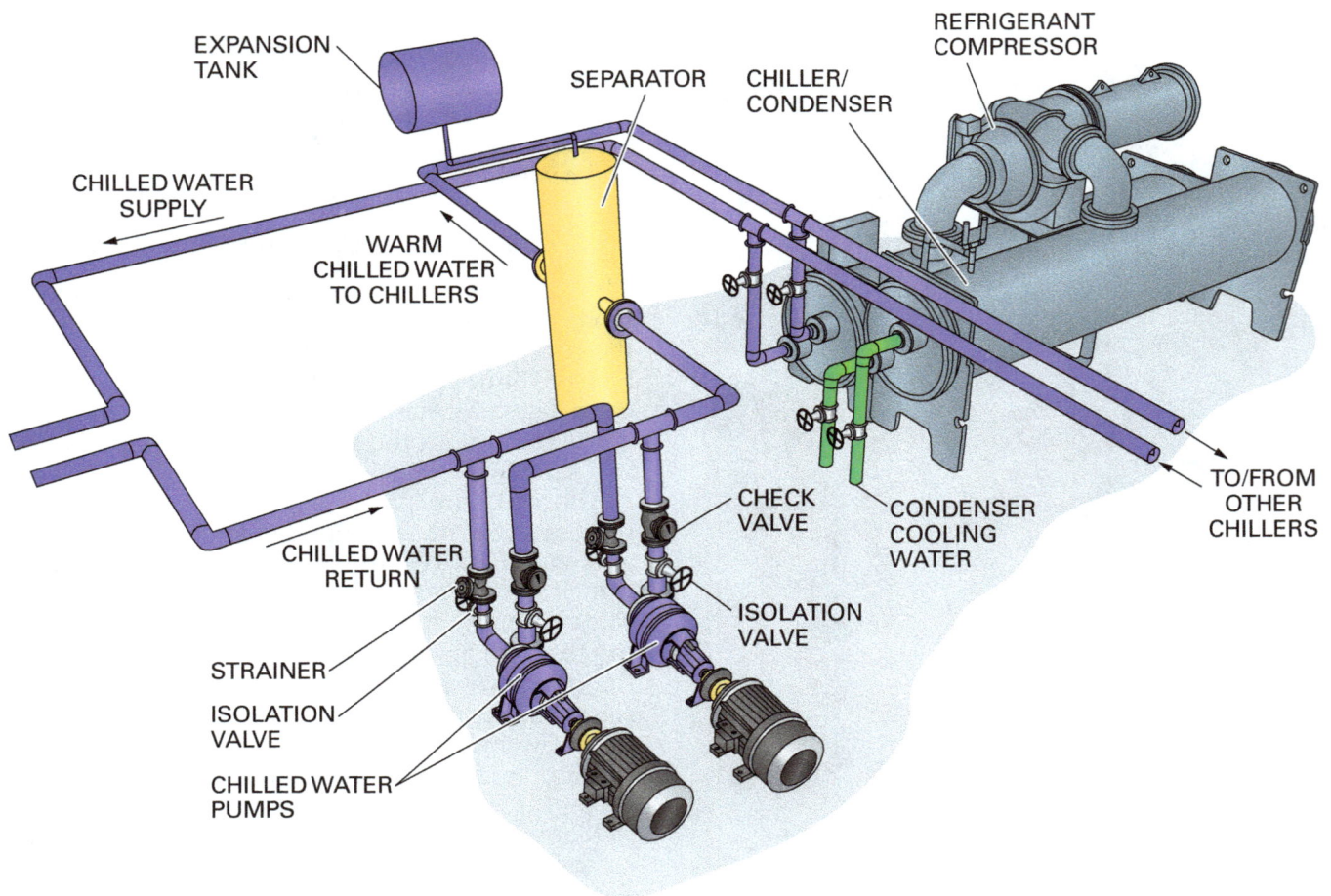

Figure 6 Chilled water plant schematic.

several cooling coils, hot water or steam heating coils, and even steam-humidifier manifolds.

Depending on the installation environment, condensate piping may also require insulation. The condensate line drains away the condensate that forms on the face of cooling coils (both chilled water and refrigerant coils), which then collects in a condensate pan under the coil. In some mechanical rooms, the temperature of the drain piping may be below ambient, causing condensation and requiring insulation (*Figure 10*).

Other chilled water piping may connect to individual room terminal units, such as a fan coil unit or induction unit. Fan coil units are small units that are located under windows, at the ceiling, or above a suspended ceiling. The unit consists of a finned-coil heat exchanger and a blower. The fan blows air over the coil, which absorbs heat from the air, cooling it. Fan coil units are available in many sizes and configurations, including the small unit shown in *Figure 11*. This particular unit is served by refrigerant piping but similar styles use chilled and/or hot water to condition the air.

An induction unit has a coil, similar to the fan coil unit, but does not have a fan. The unit is connected to a high-velocity air duct that blows air through the unit, drawing air from the space over the coil to condition the space. Both the induction unit and the fan coil unit may have hot water coils to provide heated air during the winter season.

1.1.3 Chilled Water System Configurations

Some chilled water systems may be more sophisticated than others and have primary loops furnishing chilled water to the main air handling units, with secondary loops supplying the fan coil or induction unit system (*Figure 12*). Further zoning may be done by using separate loops for different building exposures. All of this looping and zoning requires systems with intricate control valves and separate pumps.

Some systems operate chillers at night when electrical power rates are lowest to build ice in tanks. In these systems, the chilled water temperature must be reset to sub-freezing temperatures during the ice-building session. Antifreeze solutions are added to the chilled water to prevent

FLOW BALANCING VALVE

PNEUMATICALLY OPERATED FLOW CONTROL VALVE

PIPE HANGER SPRING ISOLATOR

PUMP VIBRATION ISOLATOR

Figure 7 Chilled water piping system components.

freezing internally, while ice builds on the outside of small tubes in the tank. The following day, the ice is used to chill water for building cooling while the chillers remain idle for a portion of the day. For insulators, this means that piping and component insulation must be selected and designed to accommodate pipe temperatures of roughly 20°F (–7°C) instead of typical chilled water supply temperatures.

Some industrial plants use an air washer, which cleans and cools the air, instead of an air handling unit. Air is blown through the washer where it is sprayed with chilled water. Once filtered, the water is recirculated through the chiller and sent back to spray the air stream.

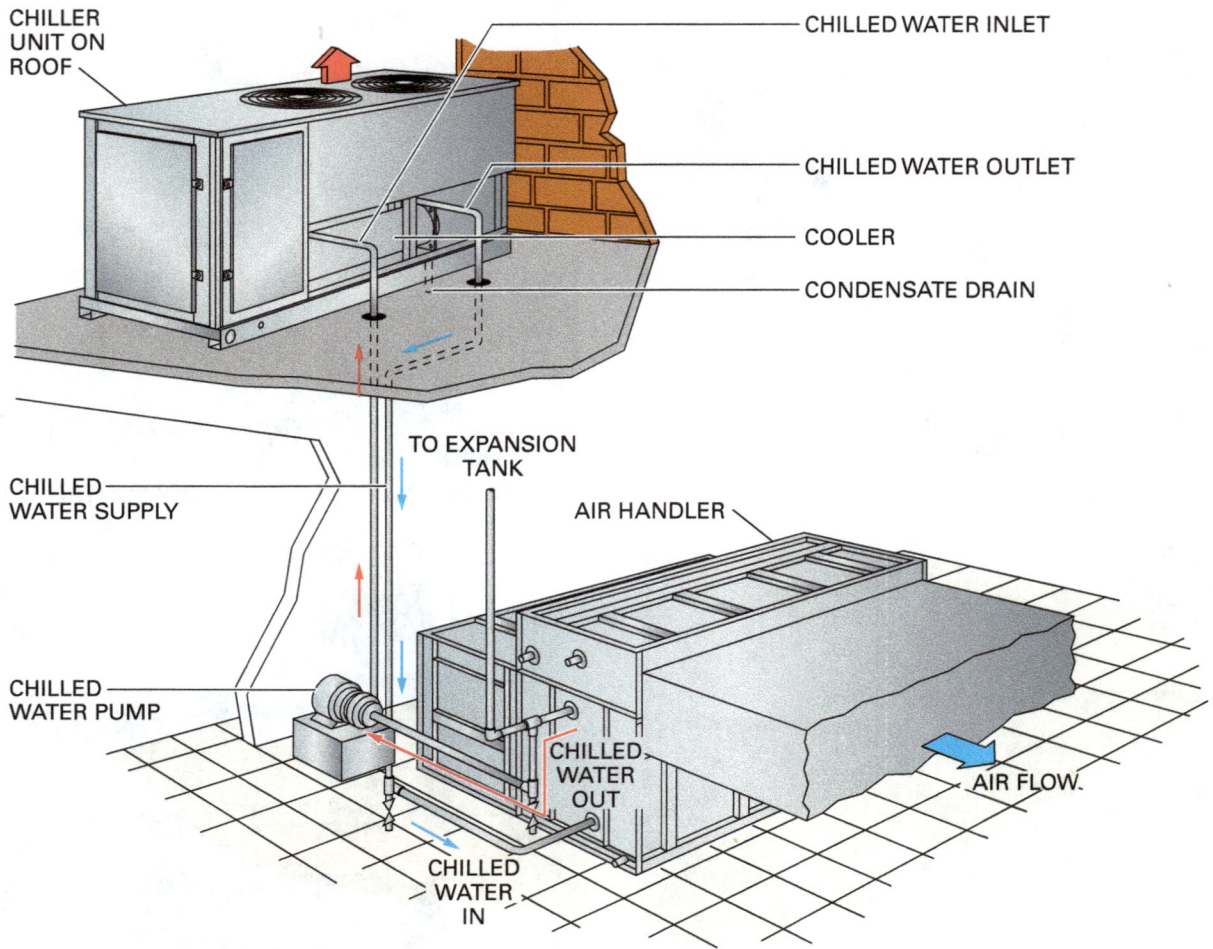

CHILLER UNIT ON ROOF

CHILLED WATER INLET

CHILLED WATER OUTLET

COOLER

CONDENSATE DRAIN

TO EXPANSION TANK

AIR HANDLER

CHILLED WATER SUPPLY

CHILLED WATER PUMP

CHILLED WATER OUT

CHILLED WATER IN

AIR FLOW

Figure 8 Chilled water system showing chiller, pump, and air handler.

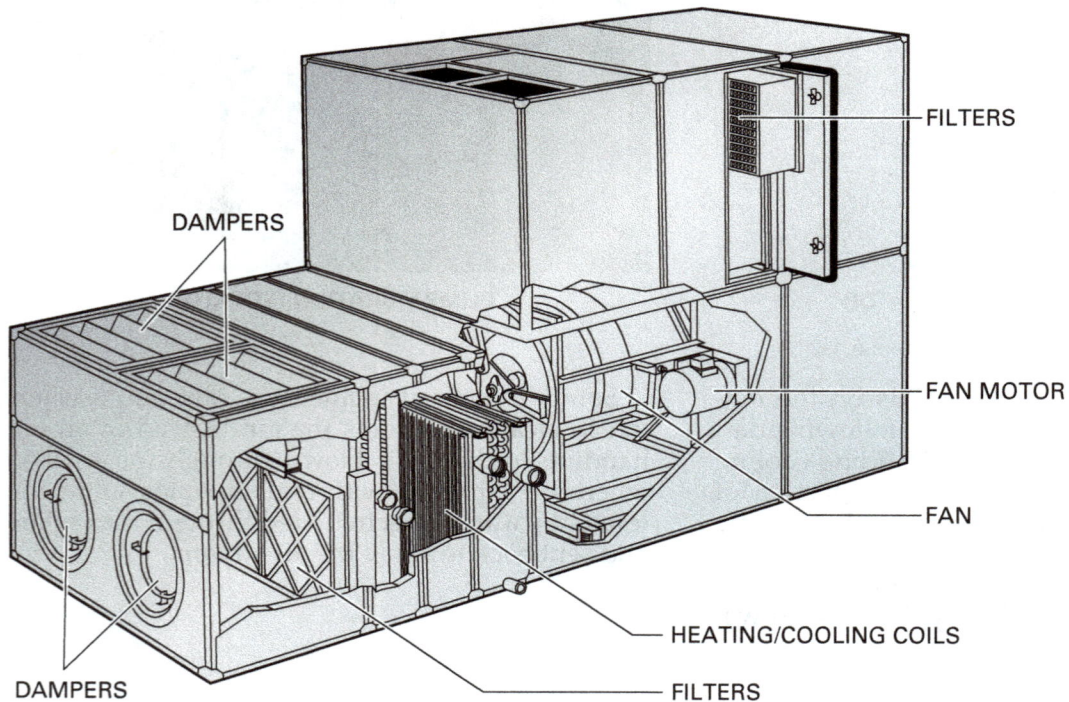

FILTERS

DAMPERS

FAN MOTOR

FAN

HEATING/COOLING COILS

DAMPERS

FILTERS

Figure 9 Commercial air handling unit.

NCCER – *Mechanical Insulating*

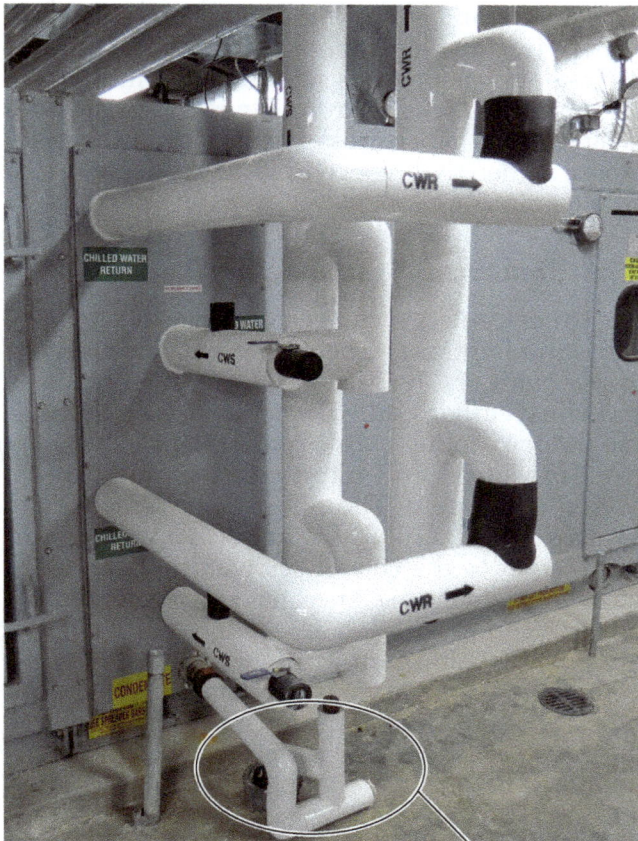

CONDENSATE
DRAIN PIPING

Figure 10 Chill water and condensate drain piping servicing
a large air handler unit.

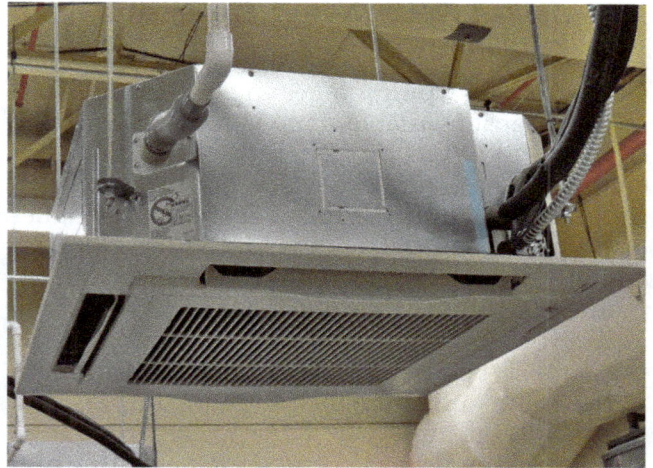

Figure 11 Fan coil unit designed for drop ceiling
installation.

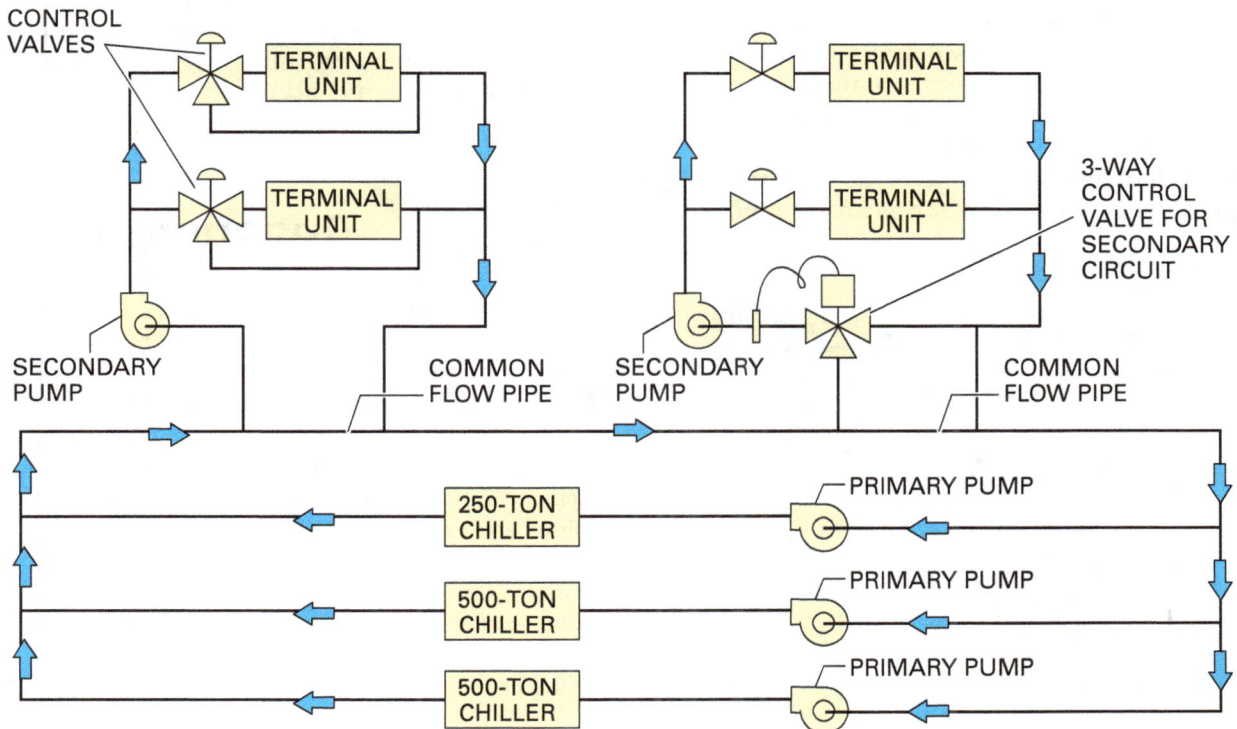

CONTROL
VALVES

TERMINAL
UNIT

TERMINAL
UNIT

SECONDARY
PUMP

COMMON
FLOW PIPE

TERMINAL
UNIT

TERMINAL
UNIT

3-WAY
CONTROL
VALVE FOR
SECONDARY
CIRCUIT

SECONDARY
PUMP

COMMON
FLOW PIPE

PRIMARY PUMP

250-TON
CHILLER

PRIMARY PUMP

500-TON
CHILLER

PRIMARY PUMP

500-TON
CHILLER

Figure 12 Simplified primary-secondary chilled water system.

1.1.4 Chilled Water System Piping and Insulation

Chilled water piping ranges in size from $\frac{1}{2}$ inch to the largest available. Pipe is made from steel, copper, and PVC. Pipes $2\frac{1}{2}$ inches and larger are commonly steel and are welded or use mechanical groove couplings. Pipes 2 inches and smaller are typically copper with soldered or solderless connections, or PVC pipes which are solvent cemented. Chilled water systems require supply and return piping operating at a relatively low pressure in a closed circuit, so structural strength is the main requirement for pipes rather than pressure resistance.

Because the water and pipe are below the ambient temperature, chilled water piping and its associated components are subject to condensation as well as heat gain and require the application of a vapor retarder. Pay close attention when vapor-sealing the insulation at valve stems, unit connections, and all breaks in the vapor retarder. Water vapor in the surrounding air will take advantage of any path through the vapor retarder and condense on the cold surfaces.

Insulation should maintain its full thickness through pipe hangers and through penetrations in walls and floors. Adequate space must be provided where piping passes close to any equipment, pipes, or walls to apply the full specified thickness of insulation and a complete vapor retarder.

To protect the vapor retarder and prevent the insulation from being crushed when suspending pipes, use heavy-density insulation supported with sheet metal shields at each hanger point. *Figure 13* shows an example of an insulated pipe supported by a clevis hanger. Note the metal shield under the insulated pipe. The weight of heavy pipe must be carried by structure as much as possible and not by the insulation material. Otherwise, the insulation will be crushed and its insulating properties significantly diminished. The metal shield shown in *Figure 11* spreads the weight of the pipe across the surface of the insulation, rather than concentrating the weight at a single point.

1.1.5 Condenser Systems

For water-cooled systems, condenser water piping is used to transfer heat collected in a building to the outdoors. It is usually large steel pipe. As described earlier, chilled water circulates through the indoor cooling coils (*Figure 11*), picks up heat from the space, and returns to the chiller to be cooled once again. The compressor removes the heat from the chilled water return piping and transfers it to the condenser water where it can be rejected outdoors. Air-cooled systems circulate hot refrigerant through an outdoor coil, where a fan typically moves air across it.

Once the condenser water has picked up the heat, it's circulated to equipment outdoors where it can be cooled. Several different types of equipment can be used. A cooling tower (*Figure 14*) is a common means of rejecting heat from the condenser water to the air around it.

Because the temperature of condenser water is the same as or warmer than room temperature, condenser water pipe requires insulation only where it is subject to freezing, i.e. outdoors. Condenser water pipe that is insulated for freeze protection is heat-traced prior to insulation. Weather protection is required over the insulation for all outdoor applications.

Figure 13 Insulated pipe in a clevis hanger.

Cogeneration

Central plant facilities that provide hot and chilled water to large service areas have long been used for college campuses, airports, high-technology manufacturing centers, and some municipalities. Central plant technology is experiencing renewed interest from energy-conscious system designers. Designers of new buildings and retrofitted high-density complexes are leaning more toward incorporating green technology into their designs. Many of these facilities now combine cogeneration capability that uses the heat byproducts of standalone electricity generation to drive the heating water system or to power chillers.

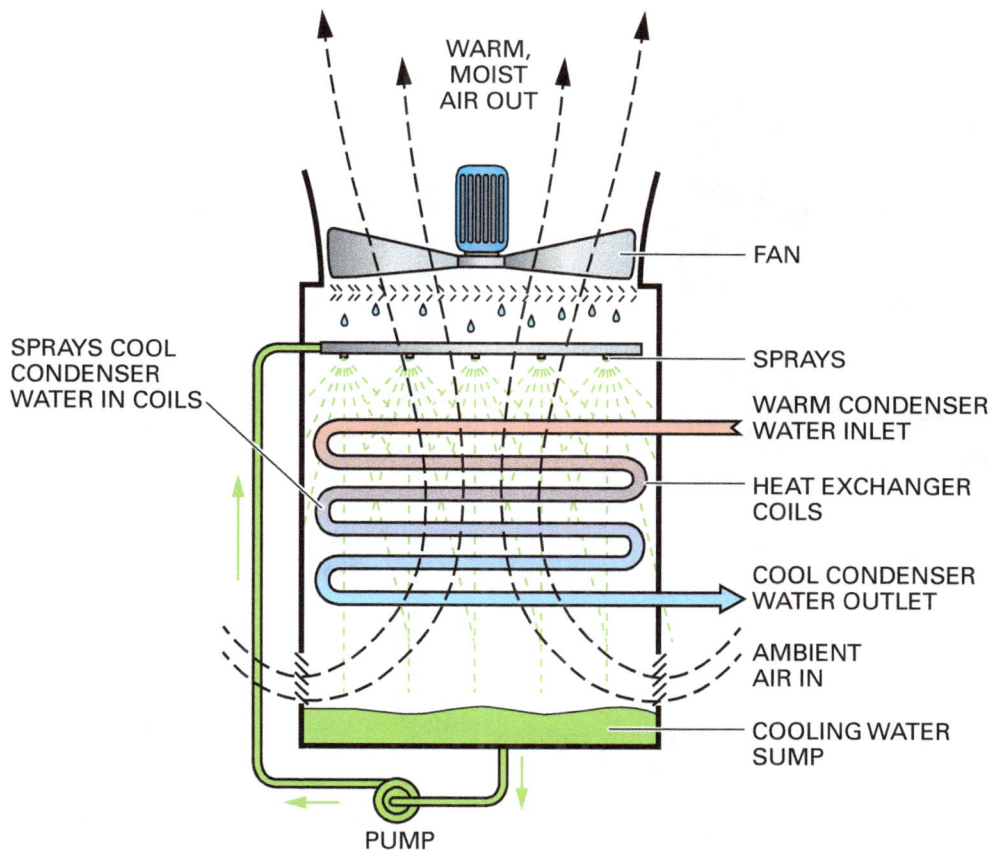

Warm, moist air out

FAN

SPRAYS COOL CONDENSER WATER IN COILS

SPRAYS

WARM CONDENSER WATER INLET

HEAT EXCHANGER COILS

COOL CONDENSER WATER OUTLET

AMBIENT AIR IN

COOLING WATER SUMP

PUMP

Figure 14 An induced-draft cooling tower.

1.2.0 Hot Water Heating Systems

Hot water heating systems are very common in commercial and industrial buildings. In a hot water heating system, the water is heated in a unit called a *boiler* (*Figure 15*). A hot water boiler can be fueled by gas, oil, coal, or electricity to directly generate the heating water. Typically, boilers for hot water heating are completely filled, while those for steam generation are designed to allow space for the steam to develop above the water level. A flame produces hot gases which pass through a series of tubes inside the boiler, trans-

ferring thermal energy to the water. Hot water piping systems require insulation to help maintain water temperature and prevent burns.

Hot water heating systems typically operate at temperatures around 170°F–190°F (77°C–88°C) and low pressures (1–2 atmospheres). They are closed-loop systems with pumps. The piping installation is similar in type and size to the chilled water system. There is, however, no concern about condensation on the pipe surface. All components of the hot water heating system are insulated only to impede heat loss from the pipe surface.

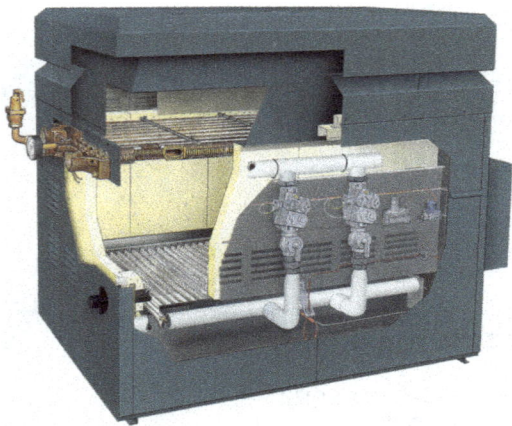

**(A) GAS-FIRED COPPER-FINNED
TUBE BOILER**

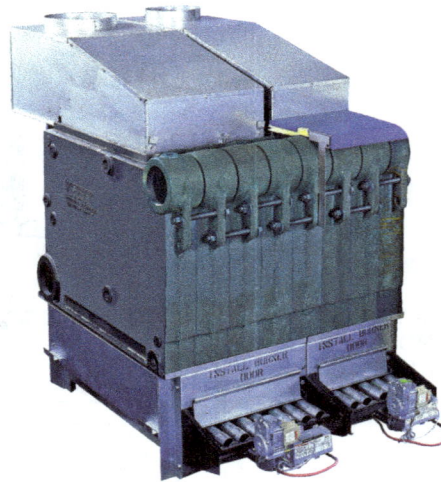

**(B) GAS-FIRED CAST IRON
SECTIONAL BOILER**

Figure 15 Typical commercial/industrial boilers.

1.2.1 Waste Steam and Steam Diversion Systems

Some industrial facilities require massive amounts of steam for production purposes. With a major supply of steam available, having separate hot water boilers for comfort purposes may not be cost-effective or reflect the wise use of resources. Steam can be diverted to a heat exchanger, also known as a *converter* (*Figure 16*). The hot water needed for comfort heating circulates through one side of the unit, while the steam circulates through the opposite side, transferring heat to the water. The heating water is then pumped to fan coil units that warm the air to transfer the heat to the space being heated.

Some systems generate high-temperature water in a central energy facility, with a distribution system connecting the various buildings. This system is similar to the chilled water system described earlier. Since the distance between the facility and the point of use may be great, the water can lose a great deal of heat even with insulation (*Figure 17*). High-temperature water for mechanical purposes is generally described as water that is heated beyond the normal boiling point (superheated) but held under pressure to prevent boiling. Water temperatures in these systems may be maintained at 450°F (232°C) or higher. When it reaches the point of use, it can be routed through a heat exchanger where heat is transferred to the building heating water loop.

> **WARNING!**
>
> High-temperature water systems can be extremely dangerous. Even the smallest pinholes or cracks that allow water to escape can cause serious bodily injury. High-temperature water escaping from a pressurized system instantly flashes to steam at atmospheric pressure. A small steam jet from a leak may be completely invisible to the naked eye, yet cut as effectively as a knife. Use extreme caution when working in and around high-temperature water systems.

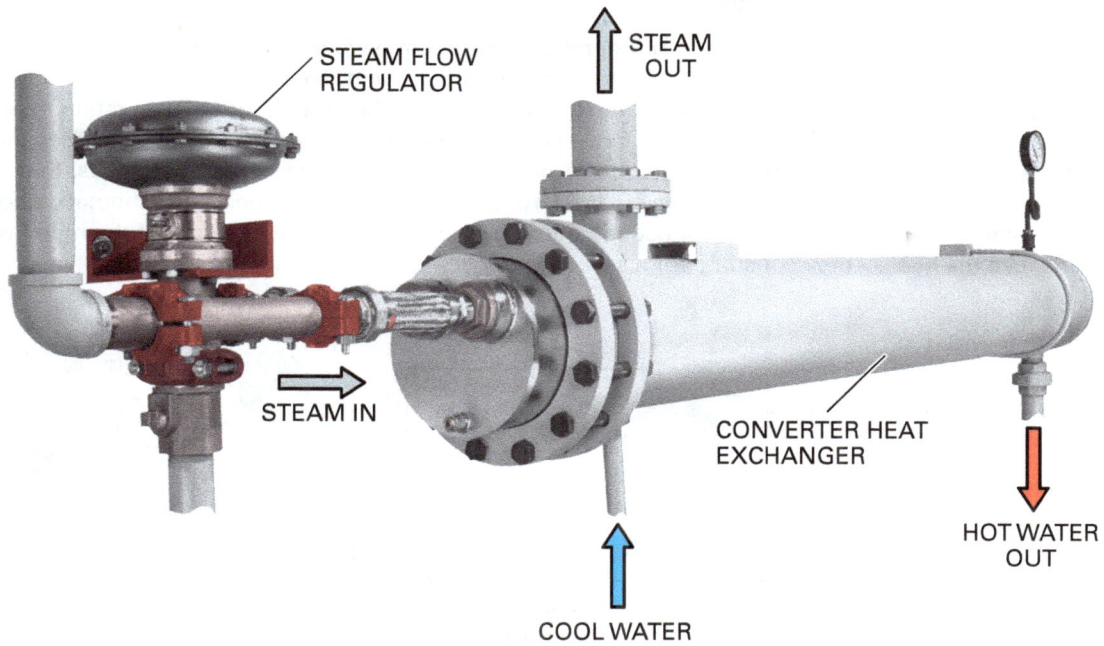

STEAM FLOW REGULATOR

STEAM OUT

STEAM IN

CONVERTER HEAT EXCHANGER

HOT WATER OUT

COOL WATER

Figure 16 Steam-to-hot water heat exchanger.

A typical commercial/industrial system will have converters and/or boilers in a mechanical room. Equipment related to cooling may share the same space, depending on the system design. As with the chilled water system, the pumps and possibly even the boiler may be duplicated to provide constant hot water heating service.

Figure 17 Insulated heating water piping.

Heat Recovery

Some air conditioning and refrigeration equipment have a heat recovery apparatus connected to a water heating system. Since refrigeration equipment, including residential air conditioning systems, are designed to transfer heat, that heat can be put to good use rather than simply rejected to the atmosphere. Through a heat recovery unit, the rejected heat, which includes the heat of compression developed in the compressor, can be used to heat or preheat water supplies. Such systems are available for residential use as well as for the largest industrial applications. The schematic shown here applies to residential and light commercial cooling equipment. This is just one example of the many ways that HVAC systems use the energy they consume for a second purpose.

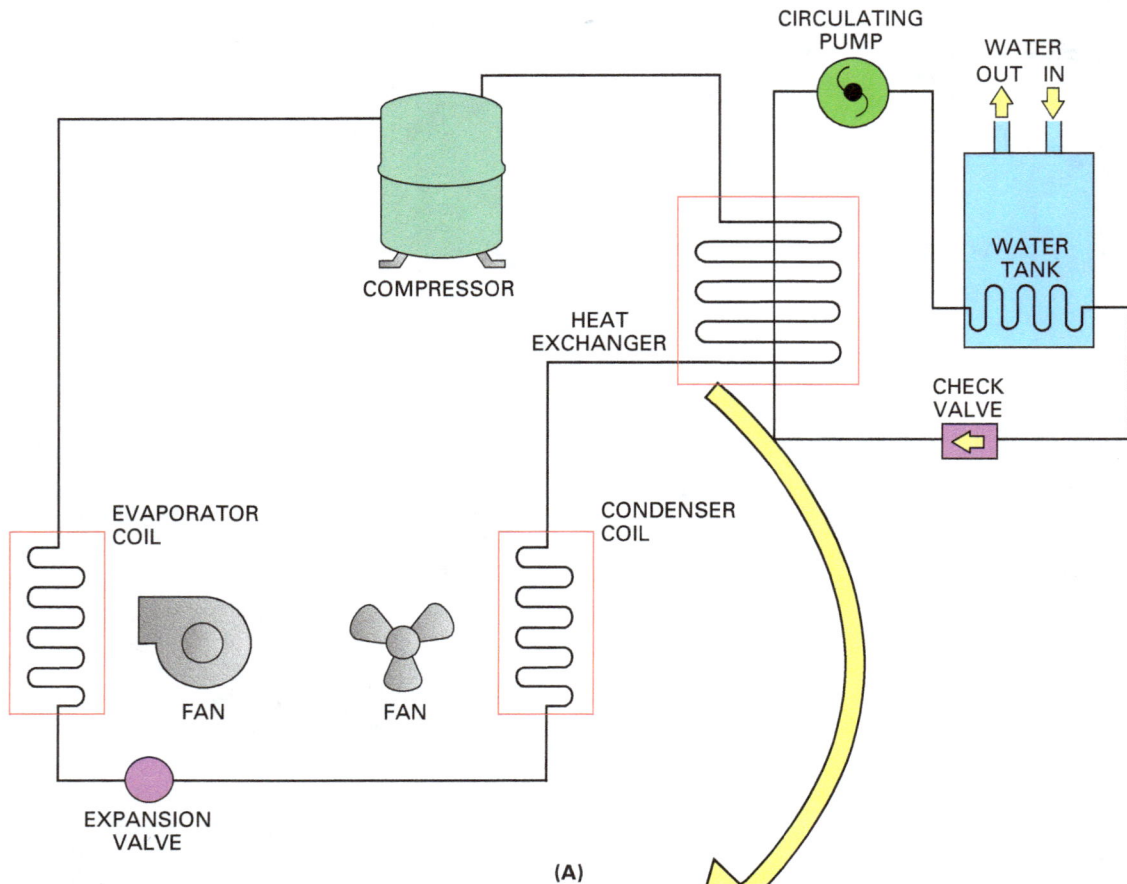

CIRCULATING PUMP

WATER OUT IN

WATER TANK

COMPRESSOR

HEAT EXCHANGER

CHECK VALVE

EVAPORATOR COIL

CONDENSER COIL

FAN

FAN

EXPANSION VALVE

(A)

ECO

(B)

1.0.0 Section Review

1. The purpose of a vapor retarder is to help _____.
    a. prevent heat loss
    b. keep chilled water cool
    c. prevent condensation
    d. prevent air loss

2. A heat exchanger is also known as a(n) _____.
    a. condenser
    b. cooling tower
    c. air handler
    d. converter

SUMMARY

Chilled water systems are safe, cost efficient ways to cool large spaces. Chillers, which may be located in a mechanical room or on the roof, cool the water. Pumps circulate the chilled water to air handlers which pass the water through coils. A fan blowing across the coil cools and distributes air through duct work to the spaces being conditioned. Instead of air handling units, some systems use air washers, which cool the air by spraying the air with the chilled water. Air separators, expansion tanks, thermometer wells, valves, chemical feed tanks, and other components may be part of the chilled water system.

Chilled water systems often have dual chillers and pumps so the system can continue to operate if one requires maintenance. The pipes in a chilled water system carry water that is below the ambient temperature. These pipes must be insulated and have a vapor retarder to prevent condensation from forming. A condenser water system circulates water, which picks up heat from the building. The warmed water is sent to cooling towers on the roof where the heat is transferred it to the air. Condenser water systems carry water that is at the ambient temperature and only need to be insulated if they are subject to freezing temperatures.

Heated water systems typically use water heated in boilers. In some large industrial facilities, steam circulates through a converter that diverts some heat from the boiler to water that will be circulated throughout the building for comfort purposes. For building heating, hot water passes through fan coils, which heat air that is sent through duct work to warm the building. Many of the piping and flow control components of a heated water system are the same as in a chilled water system. Since the water is above ambient temperature, all of the components of a heated water system must be insulated for heat retention and personnel protection. In dual systems, the pipes, air handlers, fan coils, and other components are used for both heated and chilled water. All components must be insulated for heat retention, and have a vapor retarder to prevent condensation from forming.

1. Condensation means _____.
   a. the act of getting smaller
   b. moisture from air turning into liquid
   c. liquid turning into vapor
   d. chilling warm water

2. Chilled water is _____.
   a. at ambient temperature
   b. used only in cold ambient temperatures
   c. a cooling medium in air conditioning systems
   d. not insulated

3. Why would a facility have more than one chiller?
   a. To provide lower temperatures, cooled air is run through multiple units in series.
   b. One unit would generate too much condensation.
   c. Water flow in one unit can be reversed to create heat.
   d. In case of maintenance or repair there is at least one unit to provide cooling.

4. Air conditioning chilled water is produced by _____.
   a. a cooling tower
   b. condensation unit
   c. chillers
   d. a fan

5. When does condenser water pipe require insulation?
   a. Over 100°F
   b. Freezing
   c. Dry climates
   d. Wet climates

6. Condenser water piping generally is _____.
   a. insulated
   b. made of copper
   c. small
   d. large and made of steel

7. Hot water systems require _____.
   a. a vapor retarder
   b. condensate pans
   c. insulation
   d. joint sealing

8. Steam-generating boilers are typically _____.
   a. triple-walled for safety
   b. installed on end
   c. completely filled with water to increase pressure
   d. partially filled with water to allow space for the steam

9. A converter _____.
   a. converts hot water to steam
   b. converts vapor to liquid
   c. uses steam to heat water
   d. converts liquid to vapor

10. Hot water heating systems are typically _____.
    a. open-loop systems without pumps
    b. open-loop systems with pumps
    c. closed-loop systems without pumps
    d. closed-loop systems with pumps

Trade Terms Quiz

Fill in the blank with the correct term that you learned from your study of this module.

1. Air that has been treated to control its temperature, humidity, and cleanliness to meet the requirements of a conditioned space is called _____.

2. Using a heated circulating water system to condition air within buildings in cold climates is called _____.

3. The physical process by which a vapor is converted to a liquid by cooling is called _____.

4. Water at below-ambient temperature used for cooling (particularly in air conditioning systems or in processes) is called _____.

Trade Terms

Chilled water
Condensation

Conditioned air
Hot water heating

Trade Terms Introduced in This Module

Chilled water: Water at below-ambient temperature used for cooling (particularly in air conditioning systems or in processes). Typical chilled water temperature for comfort cooling applications is 42°F–55°F (7°C–13°C).

Condensation: The physical process by which a liquid is removed from its vapor by cooling (e.g., water vapor turns into a liquid upon contact with a cold surface).

Conditioned air: Air that has been treated to control its temperature, humidity, and/or cleanliness to meet the requirements of a conditioned space.

Hot water heating: Water heated by natural or artificial means used for structural comfort heating or in processes. Typical hot water temperature for comfort heating applications is 170°F–190°F (77°C–88°C).

Additional Resources

This module presents thorough resources for task training. The following reference material is recommended for further study.

National Industrial and Commercial Insulation Standards Manual, Midwest Insulation Contractors Association (MICA).

Mechanical Insulation Design Guide, National Institute of Building Sciences. Available at **www.wbdg.org/guides-specifications/mechanical-insulation-design-guide**

National Insulation Association (NIA) website offers resources for products and training, **www.insulation.org**.

Figure Credits

Courtesy of Daikin McQuay, Module opener

McQuay International, Figure 2

Topaz Publications, Inc., Figures 3–5, 7, 10, 11

Buckaroos, Inc., Figure 13

Raypak, Inc., Figure 15A

Utica Boilers, Figure 15B

Leslie Controls Inc., Figure 16

Courtesy of ECU, SA01 (photo)

©iStockphoto.com/alacatr, Figure 17

Section Review Answer Key

Answer Section One	Section Reference	Objective
1. c	1.1.4	1a
2. d	1.2.1	1b

NCCER CURRICULA — USER UPDATE

NCCER makes every effort to keep its textbooks up-to-date and free of technical errors. We appreciate your help in this process. If you find an error, a typographical mistake, or an inaccuracy in NCCER's curricula, please fill out this form (or a photocopy), or complete the online form at **www.nccer.org/olf**. Be sure to include the exact module ID number, page number, a detailed description, and your recommended correction. Your input will be brought to the attention of the Authoring Team. Thank you for your assistance.

Instructors – If you have an idea for improving this textbook, or have found that additional materials were necessary to teach this module effectively, please let us know so that we may present your suggestions to the Authoring Team.

NCCER Product Development and Revision
13614 Progress Blvd., Alachua, FL 32615

Email: curriculum@nccer.org
Online: www.nccer.org/olf

❏ Trainee Guide   ❏ Lesson Plans   ❏ Exam   ❏ PowerPoints   Other _____

Craft / Level: _____   Copyright Date: _____

Module ID Number / Title: _____

Section Number(s): _____

Description: _____

_____

_____

_____

Recommended Correction: _____

_____

_____

_____

_____

Your Name: _____

Address: _____

_____

Email: _____   Phone: _____

This page is intentionally left blank.

Installing Fiberglass Pipe Insulation

OVERVIEW

Molded fiberglass is one of the most common insulations for pipe. Fiberglass is manufactured in forms ready-made to fit most pipe and many pipe system components. Some other pipe system components require fabrication.

Module 19106

Trainees with successful module completions may be eligible for credentialing through the NCCER Registry. To learn more, go to **www.nccer.org** or contact us at 1.888.622.3720. Our website has information on the latest product releases and training, as well as online versions of our *Cornerstone* magazine and Pearson's product catalog.

Your feedback is welcome. You may email your comments to **curriculum@nccer.org**, send general comments and inquiries to **info@nccer.org**, or fill in the User Update form at the back of this module.

This information is general in nature and intended for training purposes only. Actual performance of activities described in this manual requires compliance with all applicable operating, service, maintenance, and safety procedures under the direction of qualified personnel. References in this manual to patented or proprietary devices do not constitute a recommendation of their use.

19106 V2

INSTALLING FIBERGLASS PIPE INSULATION

Objectives

When you have completed this module, you will be able to do the following:

1. Describe the types and uses of fiberglass pipe insulation.
   a. Describe the types of fiberglass pipe insulation.
   b. Describe the uses of fiberglass pipe insulation.
   c. Describe the handling and storage requirements for fiberglass pipe insulation.
2. Explain the installation of fiberglass pipe insulation.
   a. Describe the different methods for installing fiberglass pipe insulation.
   b. Describe how to install fiberglass pipe insulation at pipe hangers, tees, and saddles.
   c. List the guidelines for a quality installation.

Performance Tasks

Under the supervision of your instructor, you should be able to do the following:

1. Apply fiberglass insulation on pipe.
2. Trim insulation and jacket.
3. Apply fiberglass insulation at hangers and supports.
4. Seal cut or punctured all-service jacketing (ASJ).

Trade Terms

Circumferential joint
Diameter
Longitudinal joint
Pipe insulation thickness
Pipe saddle

Pipe shoe
Puncture
Self-sealing lap system
Vapor retarder

Industry Recognized Credentials

If you are training through an NCCER-accredited sponsor, you may be eligible for credentials from NCCER's Registry. The ID number for this module is 19106. Note that this module may have been used in other NCCER curricula and may apply to other level completions. Contact NCCER's Registry at 888.622.3720 or go to **www.nccer.org** for more information.

Contents ────────────────────────────────

Figures ────────────────────────────────

1.0.0 FIBERGLASS PIPE INSULATION

Objective

Describe the types and uses of fiberglass pipe insulation.

a. Describe the types of fiberglass pipe insulation.
b. Describe the uses of fiberglass pipe insulation.
c. Describe the handling and storage requirements for fiberglass pipe insulation.

Trade Terms

Diameter: The distance across a circle measured through its center.

Pipe insulation thickness: The thickness of insulation to be applied to the pipe system.

Puncture: A small hole formed in an otherwise continuous material. In insulating, it tends to apply to holes made in jacket during installation.

Self-sealing lap system: An adhesive-backed strip that seals the ASJ lap (longitudinal) joint.

Vapor retarder: A material designed to minimize the passage of water vapor between two volumes of air. (All vapor retarders are permeable to vapor to some extent and should not be confused with vapor barriers.)

Before you install fiberglass pipe insulation, it must be received, stored, and moved to the proper work area. This module reviews some critical information on these topics, and includes some information specific to fiberglass pipe insulation.

1.1.0 Types of Fiberglass Insulation

Fiberglass insulation belongs to the category of fibrous insulations. It is one of the industry's most common insulations.

Fiberglass is available in several forms:

- Molded pipe insulation
- Fiberglass blanket
- Fiberglass board

When used for pipe insulation, fiberglass is compressed and molded into a shape (*Figure 1*) that fits around a standard size pipe. Fiberglass pipe insulation sections are typically 3' long. Insulation thicknesses range from ½" to 3". Greater insulation thicknesses are achieved by nesting (or

sleeving), or by special ordering a custom size from the manufacturer.

Pipe and tank wrap products are frequently used on pipe. They may be referred to as *fiberglass blanket*, although they are technically a flexible form of board insulation. Fiberglass blanket (*Figure 2*) is not typically used on pipe, although it is sometimes used to wrap larger pipes or pipe fixtures. Fiberglass blanket is presented in greater detail in Module 19202, "Blanket Insulation for Ducts" from NCCER's *Mechanical Insulating Level Two*.

Fiberglass board (*Figure 3*) is also typically not used on pipe, although there are some applications on larger sized pipe. Fiberglass board is presented in greater detail in Module 19203, "Board Insulation for Ducts" from NCCER's *Mechanical Insulating Level Two*.

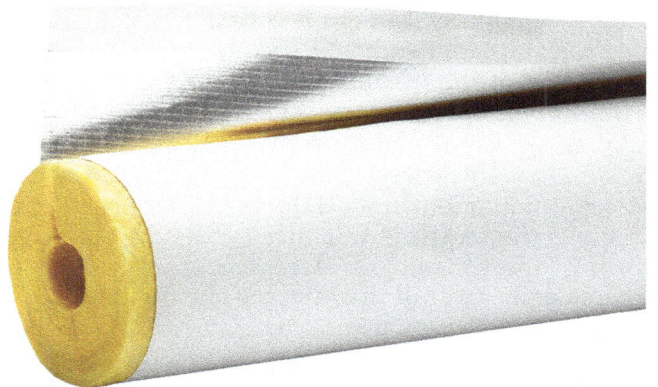

Figure 1 Molded fiberglass pipe insulation.

Figure 2 Fiberglass blanket.

Figure 3 Fiberglass board.

1.1.1 Pipe Insulation Sizing

Pipe insulation sizes are expressed in two dimensions: nominal pipe size and pipe insulation thickness. They are written using the following notation:

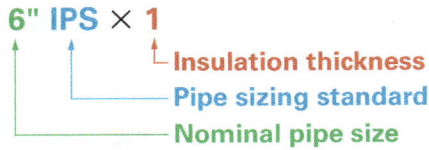

6" IPS × 1
- └ Insulation thickness
- Pipe sizing standard
- Nominal pipe size

This notation calls for insulation that is sized for 6" IPS pipe and is 1" thick.

Nominal pipe size is expressed as either an IPS or a CTS measurement. Similarly, insulation thickness is the nominal thickness of the insulating material. Note that nominal insulation thickness can vary, so a nominal 2" insulation product might be slightly more or less than 2" thick.

Project plans and specifications identify the thickness of the insulation to be applied to given equipment. While insulation is available in several thicknesses, the plan may call for a greater insulation thickness than is available. In such cases, insulation materials will be nested to achieve the required thickness. (Nesting refers to fitting larger insulation sizes over smaller ones to increase the overall thickness of the insulation.)

> **NOTE**
>
> Although insulation is manufactured to ASTM International recommended dimensions, there are allowable variations in dimensions due to manufacturing tolerances. This means that actual dimensions of materials may vary according to the type of material, the manufacturer, or even the batch. This may not generally be a concern, but when nesting a smaller product inside a larger product to achieve greater insulation thickness, tolerance variations may add up (or stack) to prevent proper fit.

Due to similarity in sizing, some IPS and CTS insulation sizes are interchangeable. Insulation sized for CTS $\frac{3}{4}$" through $1\frac{1}{2}$" can be interchanged with IPS insulation sized for a $\frac{1}{4}$" smaller nominal diameter (i.e., $\frac{1}{2}$" through $1\frac{1}{4}$" IPS). Insulation for copper pipes 2" and larger must use insulation sized for the CTS standard. It's always a good idea to check with the insulation manufacturer or fabricator to confirm which sizes are interchangeable. Smaller sizes of pipe are generally used in plumbing systems, hot water heating systems, steam pipes, chilled water pipes, and other pipe systems operating below 850°F (455°C).

1.1.2 All-Service Jacketing (ASJ)

Fiberglass pipe insulation sections are available with and without a white, low permeability, flexible reinforced foil lamination jacket historically referred to as *all-service jacketing (ASJ)*. The term ASJ includes the newest or next generation of jacketing materials, which have polymer surfaces. Each of the manufacturers has varying jacket compositions, descriptions, and brand names for the jacketing, but these are all simply referred to as *ASJ*.

The jacketing is typically supplied with a factory-applied longitudinal self-sealing lap system and furnished with matching butt strips. Some materials have a factory-applied adhesive on the contact portion of the jacket, with one protective paper between the two adhesive surfaces.

When fiberglass pipe insulation is installed on pipe systems operating at temperatures lower than the surrounding air, the ASJ helps act as a vapor retarder. The self-seal laps with the ASJ also help prevent moisture from entering the insulation. Care must be taken to make sure the ASJ is completely sealed (laps laid down tightly and any puncture sealed) to prevent moisture from entering the fiberglass. Any cuts that are made in the ASJ must be sealed using a vapor retardant mastic or tape.

1.1.3 Vapor Retarders

Vapor retarders are materials that inhibit the passage of water vapor from environments that are separated by insulation. Typically, the system being insulated is a chilled water pipe system. Water vapor that comes into contact with the system would condense to liquid water (condensation), which could damage the insulation, damage the facility, and become a safety hazard for personnel.

Several materials can be used as vapor retarders. The effectiveness of a retarder is expressed as a permeability (perm) rating, a measure of how easily water vapor is transmitted. The lower the perm rating, the better. Smaller perm ratings equate to greater resistance to the passage of water vapor. Conversely, higher perm ratings equate to less resistance to transmitting water vapor. No insulation vapor retarders are completely impermeable.

Some materials, such as vapor retardant mastic, are not applied to an entire section of insulation, but are used to seal openings in the jacket such as joints between two insulation sections or punctures.

1.2.0 Fiberglass Insulation Uses

Fiberglass pipe insulation is used on various pipe systems, connections (such as flanges), and other components such as valves. Fiberglass performs several functions, including the following:

- Keeping materials for warm/hot processes at or near their correct temperatures
- Keeping water in chilled water systems chilled
- Helping to shield chilled water pipe systems from water vapor
- Helping protect personnel from hot surfaces

1.3.0 Storage and Handling

Fiberglass pipe insulation is typically shipped to jobs in cardboard cartons. A package of butt strips is in each carton of ASJ pipe insulation, one strip for each section of insulation. The cartons may weigh from 15 to 65 lbs (7 to 30 kg) each, according to the size and thickness of the pipe insulation. Each carton is marked with the pipe size, insulation thickness, and section length of the insulation it contains. Cartons may contain as much as 300' of insulation for sizes such as $\frac{1}{2}$" IPS × $\frac{1}{2}$, and as little as 3' for sizes such as 12" IPS × 2. Fiberglass pipe insulation may be special-ordered and packaged in plastic bags.

Fiberglass pipe insulation boxes should be kept dry while in storage. The best practice is to keep boxes of insulation on wooden pallets to help keep them dry.

Fiberglass pipe insulation is semi-rigid and can be easily cut. Despite the word glass in the name, fiberglass in this form is not brittle. However, if handled or stored roughly, the ASJ will become loose, wrinkled, and dirty, and the ends of the insulation sections will become frayed and damaged. Because of this, you should handle the insulation as little as is practicable before installation.

1.0.0 Section Review

1. Fiberglass pipe insulation ranges from _____.
   a. $\frac{1}{2}$" to 3" thick
   b. 3' to 6' long
   c. 2' to 4' in width
   d. 20 to 40 pounds per foot

2. Which of the following is a use for fiberglass pipe insulation?
   a. Chilling water
   b. Heating water
   c. Providing a step surface
   d. Protecting personnel

3. A best practice for storing insulation is _____.
   a. putting it on pallets
   b. sealing it inside polyethylene bags
   c. hanging it from rafters
   d. rolling it to and from storage

2.0.0 INSTALLING FIBERGLASS PIPE INSULATION

Objective

Explain the installation of fiberglass pipe insulation.

    a. Describe the different methods for installing fiberglass pipe insulation.

    b. Describe how to install fiberglass pipe insulation at pipe hangers, tees, and saddles.

    c. List the guidelines for a quality installation.

Performance Tasks

1. Apply fiberglass insulation on pipe.
2. Trim insulation and jacket.
3. Apply fiberglass insulation at hangers and supports.
4. Seal cut or punctured all-service jacketing (ASJ).

Trade Terms

Circumferential joint: A break in the insulation encircling the pipe, formed when two sections of insulation are butted together.

Longitudinal joint: A break in the insulation formed by the split provided in molded fiberglass pipe insulation along its length on one side. The longitudinal joint is opened to slide the insulation over a pipe, then sealed after installation.

Pipe saddle: A pipe support formed from metal curved to conform to the shape of the pipe.

Pipe shoe: A pipe support made of tee iron and welded to the bottom of the pipe.

Pipe systems can contain a number of different components in addition to pipe. To be a successful mechanical insulator, you must be able to apply insulation to each type of component.

2.1.0 Installation Methods

Installing fiberglass insulation on pipe is fairly straight-forward when using the appropriate materials and tools. Careful workmanship results in an effective insulation covering and a neat installation.

2.1.1 Applying Fiberglass ASJ with Self-Sealing Laps

Fiberglass pipe insulation is slit along its length so it can be slid over the pipe. Each section of insulation is applied to a straight section of pipe, and the self-sealing lap (SSL) is used to seal the longitudinal joint.

Follow these guidelines to achieve a professional installation:

- Do not remove the SSL release paper until the insulation is on the pipe.
- Make sure the insulation is firmly closed along the slit before sealing the lap (*Figure 4*).
- Avoid compressing the insulation when sealing the lap strip. You must press firmly on the strip to ensure a good seal, but not so hard that the insulation is compressed.
- The adhesive on the self-seal contact strips works best above 60°F (16°C). Regardless of the ambient temperature, briskly rub the strip after application several times to activate the adhesive by friction, ensuring a good seal.
- If the self-seal laps fail to seal properly, secure the section using outward clinching staples in the lap. Be sure to seal all jacket punctures with vapor retardant tape or mastic.
- When a section of insulation is too long, it can be cut to the proper length with a sharp utility knife. Scissors are handy for cutting the ASJ jacket.

2.1.2 Applying Butt Strips

Each section of insulation is furnished with a butt strip. Butt strips seal the circumferential joint, the natural break in insulation where two sections of insulation butt up against each other. Follow these guidelines for a professional butt strip application:

- Make sure the two sections of insulation are butted together firmly (*Figure 5*) before starting to apply the butt strip.
- Keep the strip centered on and parallel to the butt joint.
- Cut the end of the butt strip so it is even with the longitudinal lap seal for a professional appearance.
- Press the butt strip against the insulation to ensure a good seal all the way around the insulation (*Figure 6*). Do not compress the insulation.
- If staples are required, use outward clinching staples.

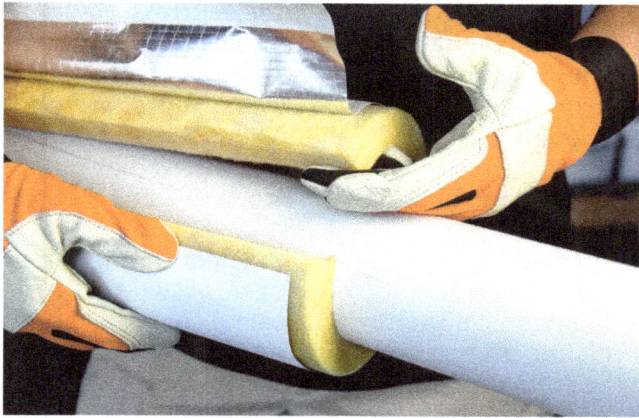

Figure 4 Applying fiberglass pipe insulation

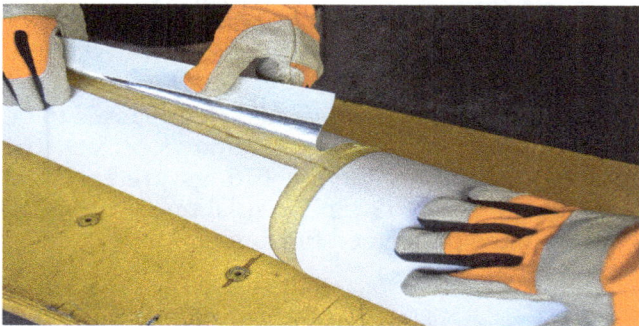

Figure 5 Insulation sections butted together.

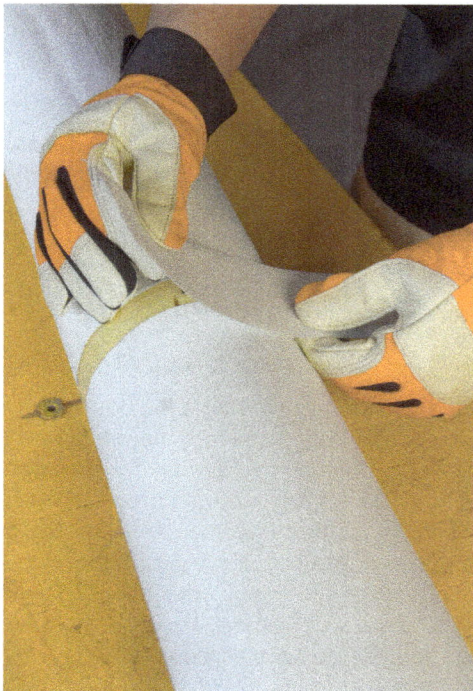

Figure 6 Applying a butt strip.

2.1.3 *Stapling and Sealing ASJ*

Always follow project specifications for applying ASJ. The following guidelines should serve as a starting point:

- Staples, when required by the specifications, should be placed on 4" centers in a straight line, no closer than ¼" from the edge of the lap.
- The staple gun should be held in loose contact with the ASJ surface. Do not force the gun into the insulation to avoid compressing the fiberglass and to avoid unnecessary breaks, deformation, or wrinkles in the jacket.
- Make sure the staple gun is flaring the staples properly; replace the staple gun head otherwise.
- Laps may be stapled to ensure additional attachment. If laps are stapled, then a vapor retardant mastic should be applied over the staples to help prevent moisture from entering the insulation at the staple penetration points.
- Plain or painted ASJ may be the final finish on pipe insulation. On the other hand, there may also be additional jackets or coatings required by the environmental or decorative requirements of the project. Note that ASJ is specifically not a weather barrier and that it cannot be installed outside a building without additional weather protection.
- When installing on cold pipe and not using the self-seal lap system, the jacket laps must be glued using a vapor retardant adhesive. Brush the adhesive on the laps after the insulation has been cut and fitted to the pipe. Allow the adhesive to become tacky and then lay the lap down and rub briskly to assure a complete seal.
- Apply a vapor retardant mastic over all staples, punctures, or cuts in the jacket if the expected operating temperature of the pipe system is below ambient.
- Some project specifications may call for jacket laps and butt strips to be coated with a vapor retardant mastic.

2.1.4 Applying Insulation Without a Jacket

Fiberglass pipe insulation can sometimes be installed without a jacket. To forego using a jacket, the insulation must not be applied to a chilled pipe system and must be installed in a concealed location. Otherwise, insulation must have a protective, vapor retardant jacketing. In such cases, the insulation can be attached with tape, wire, or bands. If the application is on a hot pipe system and the pipe is concealed in the walls or ceiling of the building, no other finish may be required. Again, check with the insulation manufacturer for recommended practices applying to the case in question.

> **CAUTION**
>
> When using wire to secure bare fiberglass insulation, be sure that the wire isn't tight enough to cut or compress the insulation.

Remember that, while fiberglass insulation is rigid enough to maintain its own shape, it can't carry loads. Fiberglass insulation needs to be adequately protected where it may become compressed (for example, in a location where foot traffic is to be expected, such as near a ground level pipe rack).

2.1.5 Trimming Insulation

As you install pipe insulation, you will encounter obstructions. Some of these obstructions require special fitments to be constructed to enclose them. Others, such as weld-on threaded and smooth bosses (*Figure 7*), require trimming the insulation to fit around them and the connected instrumentation or piping. Couplings and tees on copper pipe also typically require some trimming. Follow these guidelines to trim insulation properly:

THREDOLET® WELDED BOSSES

Figure 7 Threaded bosses welded on a pipe.

- As always, use a sharp knife to trim a cavity in insulation that will allow the insulation to fit snugly against the obstruction.
- Trim the insulation in layers, removing only enough to allow for the insulation to fit snugly against the fitting. You can always remove more insulation, but once trimmed it can't be untrimmed.
- Insulation is usually stopped at or butting up to pipe fittings.
- When installing fiberglass on a cold pipe application, a vapor retardant seal is sometimes applied to the trimmed end of the insulation, often a mastic which forms a seal from the jacket to the pipe.
- Where pipe insulation runs up to a bolted pipe flange, the insulation is either beveled or stopped with sufficient room to allow the removal of the flange bolts for servicing, typically the length of the bolt plus 1" (*Figure 8*).
- If preformed or prefabricated fiberglass is to be used on 90- and 45-degree elbows or tees, the insulation pieces for the fittings are installed first and then the pipe insulation is cut to fit against the preformed fittings.

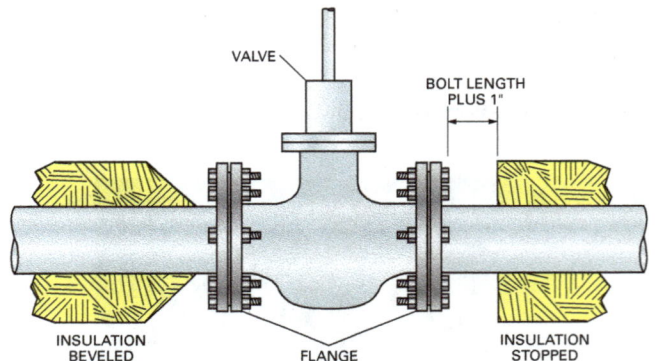

Figure 8 Stopped and beveled insulation sections near a pipe flange.

Did You Know?

Threaded and smooth welded pipe bosses called Thredolets® and Weldolets® are manufactured by Bonney Forge®, which owns those trademarks. Other similar items are manufactured by other companies. In common use you will hear them described using Bonney Forge's trademarked terms. All of these types of items work similarly. The fitting is welded to the pipe, usually before the pipe is in use, although some specialized equipment can be used to install them on operating pipe. In those cases, the pipe is punctured to permit flow through the fitting.

2.2.0 Fiberglass Insulation on Contact-Type Hangers

Hangers can either go inside or outside the insulation; that is, hangers can either support the pipe directly or support the insulation and the pipe together.

Follow these guidelines for a professional application around contact-type hangers and supports (*Figure 9*):

Before measuring or trimming insulation, make sure the hanger is straight and plumb on the pipe.

- Make sure that the holes for the hanger rod in both the insulation and any jacket are cut so they can line up.
- Remove only enough insulation for a snug fit around the hanger.
- Fill any voids between the pipe and the hanger with scrap insulation.
- Apply a vapor retardant around the hanger hole.

2.2.1 Installing Fiberglass Insulation Using Rigid Inserts and Sheet Metal Shields

If a pipe hanger is to be outside the insulation, there is a danger that the weight of the pipe will compress the insulation, reducing its effectiveness. In such cases, the pipe needs to be supported by either a rigid insert (*Figure 10*) beneath it, a sheet metal shield outside the insulation (*Figure 11*), or both.

> **NOTE**
>
> Most cold pipe applications use hangers outside the insulation.

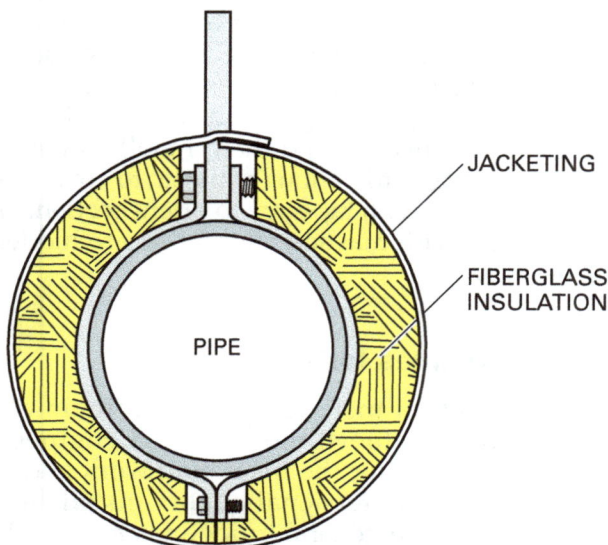

Figure 9 Contact-type hanger.

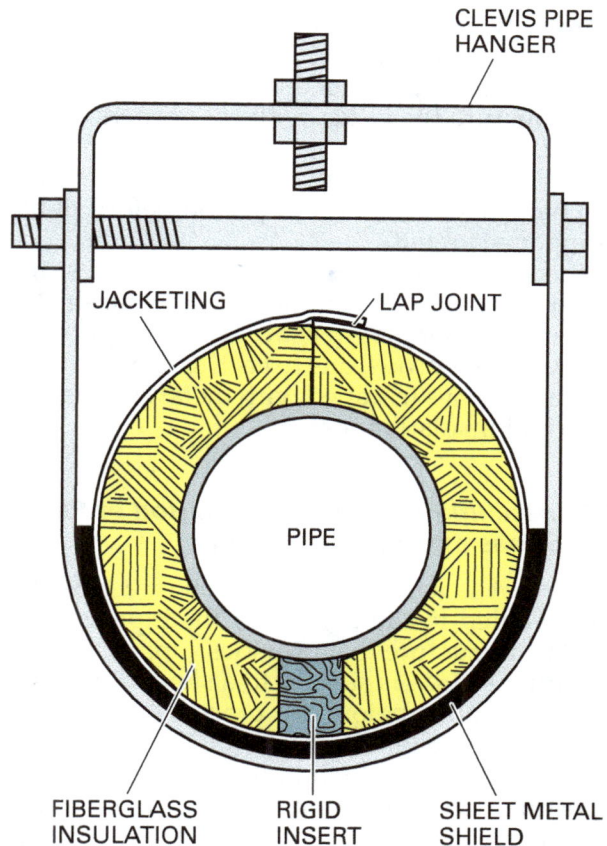

Figure 10 Weightbearing rigid insert.

Figure 11 Sheet metal shield.

Follow these guidelines to apply insulation where the hanger is outside the insulation:

- Never loosen more than one hanger at a time. Loosening two or more hangers at one time can allow the pipe to sag and cause damage.
- If you can arrange it, have the pipefitters install the hangers high on the hanger rod. When it's time to apply the insulation you will have to lower the hanger; if it's already at the bottom of the rod there will be no room to do so.

- Trim only as much insulation as necessary for a snug fit around a rigid insert.
- When installing a metal shield, be careful to not puncture the jacket with the corners of the shield.
- After resetting the hanger height, make sure the hanger is straight and plumb before tightening the jam nut.
- If the project requires additional finishes, install them before installing a sheet metal shield.
- Roller-type hangers (*Figure 12*) are supported by two hanger rods, one to each side of the pipe and are always installed with metal shields under rigid insulation inserts.

2.2.2 Installing Fiberglass Insulation at Pipe Shoes and Saddle Mounts

A pipe shoe is a section of tee iron or a box-like fitting welded to the bottom of the pipe and attached to a structure that provides support (*Figure 13*). A pipe saddle is similar to a shoe. It consists of a curved plate below the pipe that conforms to the pipe's outside diameter (OD). The plate is welded to a stand, which is then connected to a supporting structure (*Figure 14*). A saddle may include a clamping ring over the top of the pipe, or it may be welded to the pipe.

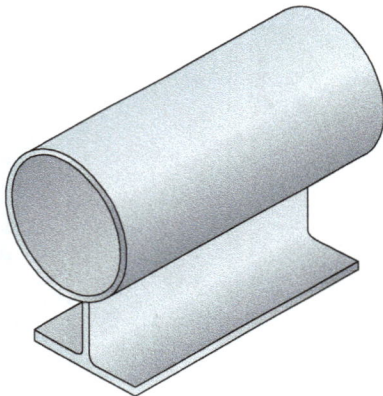

Figure 13 Pipe shoe.

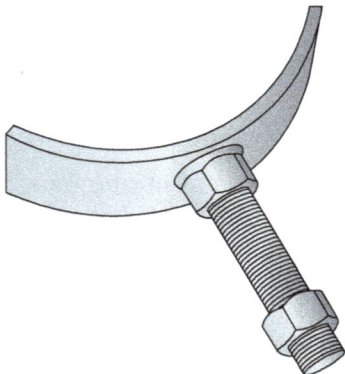

Figure 14 Pipe saddle.

Figure 12 Roller-type support.

Shoes and saddles are normally used on hot pipe systems. Follow these guidelines to apply insulation around shoes and saddles.

- Insulation should be trimmed for a snug fit around the shoe or saddle. One strategy is to use cardboard to make a cutting template.
- Use the trimmed insulation to fill any voids.
- Seal any gaps in jacketed insulation with vapor retardant.

2.2.3 Sealing Cut or Punctured ASJ

In general you should avoid puncturing or cutting ASJ. Some applications do require the ASJ to be trimmed for proper fitting, however. If staples are required for the installation, the ASJ will be punctured by the staples. ASJ may also be accidentally punctured during installation.

When applying fiberglass to chilled water systems, any openings in the ASJ must be sealed with a vapor retarder such as tape or mastic. Follow these guidelines:

- Tape can be applied directly over a puncture or cutout. Tape should be applied either horizontally or vertically, not at an angle. If possible, the taped section should be rotated to face a wall or oriented so it is otherwise hidden.
- Mastic should be applied in a layer slightly thicker than needed and then smoothed with a trowel. Apply only enough mastic to cover the puncture. As with tape, try to orient sections of insulation with mastic seals so that are hidden from view.

2.3.0 Guidelines for a Quality Job

As an insulation installer, your two main objectives are to achieve the designed engineering specifications for the insulation work and to produce a neat, professional-quality installation. The following are general guidelines for installing and working with fiberglass pipe insulation.

Following these guidelines will ensure a quality result:

- When using a cut section of pipe insulation, apply the remainder of the cut insulation section to the other side of the hanger to prevent waste.
- After completing knife or saw cuts on any fiberglass pipe covering, trim the jacket with scissors for a neat appearance.
- On cold applications, the branch line junction-to-main line jacket should be sealed using a vapor retardant mastic.
- When taking measurements for cuts in a pipe covering, make measurements in the direction toward the cut end. That is, measure from the last piece of insulation to the hanger and not from the hanger to the insulation. This helps prevent laps from being turned the wrong direction.

- All materials must be stored properly, kept dry, and protected from physical damage. It is a good practice to place materials on pallets or boards to keep them off the floor.
- Boxes should not be opened until material is to be used. A box should be completely used before another box is opened.
- When fiberglass is snagged or pulled from the insulation section, the jacket becomes wrinkled, dirty, and loose. The ends become frayed and damaged. This makes for poor quality when these materials are installed. Materials should always be protected and kept neat.
- At the end of the work shift, cartons of material should be stacked in an orderly fashion. They should be moved from aisles or passageways. Ladders and equipment must be secured. On some jobs, such as in plants that are in operation, all material and equipment must be returned to the main storage area at the end of each day.

2.0.0 Section Review

1. Why should a staple gun not be pressed into insulation when stapling ASJ?
   a. The staple might pass through the insulation.
   b. The insulation might be compressed or damaged.
   c. The staple might damage the pipe.
   d. The density of the ASJ might cause the staple gun to jam.

2. Sheet metal shields should be installed carefully to prevent _____.
   a. tearing the ASJ
   b. discoloring the ASJ
   c. twisting the shield
   d. misaligning the shield

3. As an insulation installer, your two main objectives are to achieve the designed engineering specifications for the insulation work and to _____.
   a. keep ready-to-use insulation out of the way of traffic
   b. prevent physical damage to stored insulation
   c. avoid wasting insulation
   d. produce a neat, professional-quality installation

SUMMARY

This module covers one of the most important topics for insulation trainees. Fiberglass pipe insulation is one of the most commonly used materials in the insulation trade. The skills and knowledge needed to apply fiberglass pipe insulation are applicable to some other types of insulations as well.

Remember that a good installation begins with good storage and proper handling. Be sure that insulation is kept dry and clean in order to make a professional job easier.

While several skills applicable to fiberglass pipe insulation are also applicable to additional pipe system components, other components require different skills. You have learned many of those skills in this module, and more skills will be presented in subsequent modules.

Because this material is crucial to the mechanical insulator, make sure you have a thorough understanding of it. Practice until you are comfortable with all Performance Tasks. You will use these skills throughout your career.

1. Fiberglass pipe insulation sections are typically_____.
   a. $\frac{1}{8}$" long
   b. $\frac{1}{2}$" long
   c. 1' long
   d. 3' long

2. When describing pipe insulation size, which of the following is stated first?
   a. Insulation thickness
   b. Pipe size
   c. Jacket type
   d. Sections per carton

3. Sections of ASJ insulation are typically sealed with _____.
   a. self-sealing laps
   b. cyanoacrylate adhesive
   c. wood glue
   d. twisted wire

4. When acting as a vapor retarder, ASJ must _____.
   a. be installed with no open punctures, gaps, or cuts
   b. be colored white
   c. be metallic
   d. cover all components of a piping system

5. The insulation for a chilled water system _____.
   a. does not require a vapor retarder
   b. should have a high perm rating
   c. should have a low perm rating
   d. requires metal sheathing

6. Which of the following is a purpose of fiberglass pipe insulation?
   a. Maintaining materials at or near their correct temperatures.
   b. Preventing high-pressure damage to pipe.
   c. Maintaining pipe system pressure.
   d. Protecting equipment from damage.

7. If not handled and stored properly, ASJ will _____.
   a. break in jagged pieces
   b. become loose and wrinkled
   c. disintegrate over time
   d. shrink significantly

8. Insulation in storage should be kept _____.
   a. dry
   b. compressed
   c. above eye level
   d. wrapped in clear plastic

9. Self-sealing lap release paper should not be removed until _____.
   a. you have removed your work gloves
   b. the insulation is in storage
   c. the temperature is over 68°F (20°C)
   d. the insulation is applied to the pipe

10. Butt strips are applied _____.
    a. parallel to the butt joint
    b. perpendicular to the butt joint
    c. in concentric rings
    d. over one another

11. How might you reduce the effectiveness of insulation when installing a butt strip?
    a. Allow the adhesive to get caught on the insulation.
    b. Compress the insulation by pressing too firmly to seal the butt strip.
    c. Seal the butt joint too tightly to breathe.
    d. Apply the butt strip dry, so that it doesn't seal.

12. When stapling ASJ, the staples should be _____.
    a. inward-clinching
    b. sealed with a butt strip
    c. no closer than 6" from another staple
    d. no closer than $\frac{1}{4}$" from the edge of the lap

13. Which of these applications may not require jacketed insulation?
    a. A chilled water pipe system.
    b. A plastic pipe system.
    c. A pipe system on which the insulation is concealed.
    d. Pipe system temperature above 100°F (38°C).

14. If project specifications call for an additional pipe finishing at a location supported by a sheet metal shield, that finish should be installed _____.
    a. first
    b. last
    c. before the pipe is installed
    d. before the sheet metal shield

15. At the end of a work shift, ladders should be _____.
    a. hung from hooks
    b. taken back to your workplace
    c. secured in locked storage
    d. placed against an out-of-the-way wall

Trade Terms Quiz

Fill in the blank with the correct term that you learned from your study of this module.

1. A pipe support fabricated from tee iron is called a _____.

2. The _____ seals the length-wise slit in the insulation after it is installed on a pipe.

3. A material that is used to help lessen the likelihood of condensation is called a _____.

4. The slit that runs along the length of a section of pipe insulation is called the _____.

5. A _____ is a pipe support that conforms to the shape and external size of the pipe.

6. Any _____ in ASJ must be sealed to help prevent condensation.

7. The break in the insulation created where two sections of insulation butt together is called the _____.

8. The _____ of a circle is the distance across the circle through its center point.

9. The distance from the inner surface of a section of applied insulation to the outer surface is called the _____.

Trade Terms

Circumferential joint
Diameter
Longitudinal joint

Pipe insulation thickness
Pipe saddle
Pipe shoe

Puncture
Self-sealing lap system
Vapor retarder

Trade Terms Introduced in This Module

Circumferential joint: A break in the insulation encircling the pipe, formed when two sections of insulation are butted together.

Diameter: The distance across a circle measured through its center.

Longitudinal joint: A break in the insulation formed by the split provided in molded fiberglass pipe insulation along its length on one side. The longitudinal joint is opened to slide the insulation over a pipe, then sealed after installation.

Pipe insulation thickness: The thickness of insulation to be applied to the pipe system.

Pipe saddle: A pipe support formed from metal curved to conform to the shape of the pipe.

Pipe shoe: A pipe support made of tee iron and welded to the bottom of the pipe.

Puncture: A small hole formed in an otherwise continuous material. In insulating, it tends to apply to holes made in jacket during installation.

Self-sealing lap system: An adhesive-backed strip that seals the ASJ lap (longitudinal) joint.

Vapor retarder: A material designed to minimize the passage of water vapor between two volumes of air. Vapor retarders are sometimes (incorrectly) referred to as vapor barriers, but all vapor retarders are permeable to vapor to some extent.

Figure Credits

Topaz Publications, Inc., Module opener
Photos courtesy of Owens Corning, Figures 1–3
Ron Yoakum, Figure 7
Buckaroos, Inc., Figure 11
Anvil International, Inc. Figure 12

Section Review Answer Key

Answer	Section Reference	Objective
Section One		
1. a	1.1.0	1a
2. d	1.2.0	1b
3. a	1.3.0	1c
Section Two		
1. b	2.1.3	2a
2. a	2.2.1	2b
3. d	2.3.0	2c

NCCER CURRICULA — USER UPDATE

NCCER makes every effort to keep its textbooks up-to-date and free of technical errors. We appreciate your help in this process. If you find an error, a typographical mistake, or an inaccuracy in NCCER's curricula, please fill out this form (or a photocopy), or complete the online form at **www.nccer.org/olf**. Be sure to include the exact module ID number, page number, a detailed description, and your recommended correction. Your input will be brought to the attention of the Authoring Team. Thank you for your assistance.

Instructors – If you have an idea for improving this textbook, or have found that additional materials were necessary to teach this module effectively, please let us know so that we may present your suggestions to the Authoring Team.

NCCER Product Development and Revision
13614 Progress Blvd., Alachua, FL 32615

Email: curriculum@nccer.org
Online: www.nccer.org/olf

❏ Trainee Guide ❏ Lesson Plans ❏ Exam ❏ PowerPoints Other _____

Craft / Level: _____ Copyright Date: _____

Module ID Number / Title: _____

Section Number(s): _____

Description: _____

_____

_____

_____

_____

Recommended Correction: _____

_____

_____

_____

_____

Your Name: _____

Address: _____

_____

Email: _____ Phone: _____

This page is intentionally left blank.

Insulating Pipe Fittings, Valves, and Flanges

OVERVIEW

All parts of a pipe system must be insulated. While applying insulation to a straight run of pipe is relatively simple, it is likely that most of your time will be spent forming insulation covers for fittings and valves. This can be very difficult and requires a great deal of skill.

Module 19107

Trainees with successful module completions may be eligible for credentialing through the NCCER Registry. To learn more, go to **www.nccer.org** or contact us at 1.888.622.3720. Our website has information on the latest product releases and training, as well as online versions of our *Cornerstone* magazine and Pearson's product catalog.

Your feedback is welcome. You may email your comments to **curriculum@nccer.org**, send general comments and inquiries to **info@nccer.org**, or fill in the User Update form at the back of this module.

This information is general in nature and intended for training purposes only. Actual performance of activities described in this manual requires compliance with all applicable operating, service, maintenance, and safety procedures under the direction of qualified personnel. References in this manual to patented or proprietary devices do not constitute a recommendation of their use.

19107 V2

INSULATING PIPE FITTINGS, VALVES, AND FLANGES

Objectives

When you have completed this module, you will be able to do the following:

1. Describe the different types of valves and their insulation requirements.
   a. Describe the different types of valves.
   b. Explain the insulation practices related to non-flanged valves.
   c. Explain the insulation practices related to flanged valves.
2. Describe the different types of fittings and joints and their insulation requirements.
   a. Describe the different types of fittings and joints.
   b. Explain the insulation practices related to non-flanged fittings and joints.
   c. Explain the insulation practices related to flange fittings.
   d. Explain the insulation practices related to mechanical fittings.

Performance Tasks

Under the supervision of your instructor, you should be able to do the following:

1. Apply insulation to non-flanged valves.
2. Apply insulation to flanged valves.
3. Apply insulation to flanged fittings.
4. Apply insulation to non-flanged fittings.
5. Apply insulation to mechanical fittings.

Trade Terms

Bonnet
Heel
Hydraulic setting insulation cement
Long-radius elbow
Miters

Packing gland
Radius
Short-radius elbow
Throat

Industry Recognized Credentials

If you are training through an NCCER-accredited sponsor, you may be eligible for credentials from NCCER's Registry. The ID number for this module is 19107. Note that this module may have been used in other NCCER curricula and may apply to other level completions. Contact NCCER's Registry at 888.622.3720 or go to **www.nccer.org** for more information.

Contents

Figures and Tables

This page is intentionally left blank.

1.0.0 VALVES

Objective

Describe the different types of valves and their insulation requirements.

   a. Describe the different types of valves.
   b. Explain the insulation practices related to non-flanged valves.
   c. Explain the insulation practices related to flanged valves.

Performance Tasks

   1. Apply insulation to non-flanged valves.
   2. Apply insulation to flanged valves.

Trade Terms

Bonnet: A cover for a valve body that can be removed to allow for valve maintenance and repair.

Packing gland: An assembly that houses a valve and contains a sealing material (packing), used to form a complete seal and prevent leakage.

Valves are used to control, regulate, or maintain a characteristic of a pipe system, such as pressure, temperature, or flow. Valves are available in a variety of materials, each made specifically for one or more types of pipe.

1.1.0 Valve Types

There is a wide array of valves available for pipe systems. Valve designs found in a pipe system vary according to the following:

- The purpose of the fitting or valve
- The material used in the pipe system
- The attachment method used
- The pipe system operating pressures

For example, copper pipe and fittings are frequently used in plumbing water applications. High-pressure steam pipe systems, on the other hand, call for welded steel pipe and fittings.

Valves that are operated to exert control over a characteristic of a pipe system by opening or closing, either fully or partially, are called *control valves*. Usually, they are automatically operated by an outside source of energy, which can be electric, hydraulic, or pneumatic, etc. They may control the flow of material, the temperature of the outputs or inputs of a material process, or the pressure. There are many types of control valves for different purposes, such as gate valves that are used to stop flow.

Some valves are used to prevent fluid flow in the wrong direction or to relieve excess pressure in the pipe system. These are not typically controlled or positioned by outside sources of energy. Examples include check valves and relief valves.

Common types of valves include the following:

- *Gate valve* – Operates by moving a gate in or out of the path of material flow. Gate valves come in all sizes and are used to stop flow, rather than throttle or regulate flow volume. A large flanged gate valve is shown in *Figure 1*.
- *Mixing and diverting valves* – These are three-way valves with different functions. A mixing valve accepts two separate streams of fluid and mixes them together into a single stream leaving the valve. A diverting valve accepts one fluid stream in and diverts it to one of two possible outlets.
- *Butterfly valve* – Similar to a gate valve, a butterfly valve works by rotating a disc across the interior of the valve body.
- *Balancing valve* – Used to maintain the flow of fluid at a specific volume. For example, the flow of hot water to a heating coil should be at the design volume. Too little means not enough heat; too much may cause other coils connected to the same system to starve. A balancing valve ensures that the flow volume remains consistent, regardless of other system conditions.
- *Check valve* – Allows fluid to flow in only one direction. Simple check valves depend on gravity or fluid movement in the wrong direction to close. More complex designs incorporate springs that must be overcome by the fluid flow in the correct direction to open. A ball check valve and a simple swing check valve that depends on gravity for closure are shown in *Figure 2*.
- *Relief valve* – Like control valves, relief valves (*Figure 3*) are usually automated. However, they very rarely use an external source of energy, since pressure relief may be required even when no power is present to prevent the rupture of a tank or piping system. They act as a safety to release excess pressure or excess heat before systems become damaged.

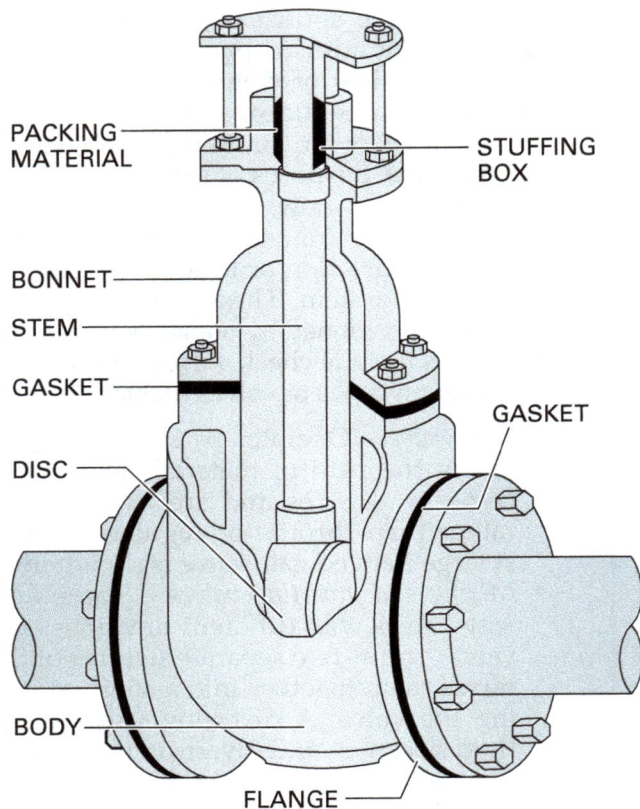

Figure 1 Gate valve.

1.2.0 Insulating Non-Flanged Valves

Non-flanged valves are relatively common on smaller pipe. Flanges are generally used on 2" (DN50) pipe or larger. They may be used on smaller sizes of pipe in unusual applications. Threaded fittings, however, are far more common on pipe sizes 2" (DN50) and under. Follow these guidelines when insulating non-flanged valves:

- If the project specifications allow the insulation to be thinned, trim out a space for the fitting to fit snugly, with the valve handle protruding.
- If insulation thinning is not allowed, wrap the pipe with flexible fibrous insulation, then use an oversized pipe insulation that nests over the flexible insulation.
- Non-flanged valves can be insulated with the same methods as flanged valves, except for removable, reusable flexible insulation covers.

1.3.0 Insulating Flanged Valves

The flanges on flanged valves are the same size and shape as those used for any flanged fitting or connection for a given size of pipe. Follow these guidelines for a professional installation:

- Installation of insulation on a flanged valve will require cutouts to clear the valve bonnet.
- Insulation for the flanged valve should extend past the flange bolts (not the flange body) for a distance equal to the insulation thickness on each end.
- Make sure the valve body insulation is level and plumb. True-up the exposed sections.
- If the bonnet flange is oval, insulate with pipe insulation over the rounded ends and block or board insulation between the rounded ends.
- Do not cover the packing gland bolt heads when insulating the valve body.
- Fill voids with scrap insulation.
- Vapor-seal all cold pipe applications.

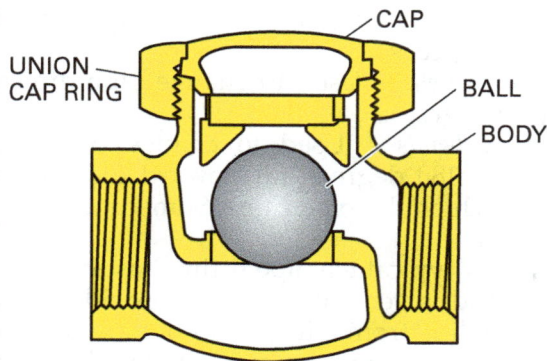

(A) BALL-CHECK VALVE

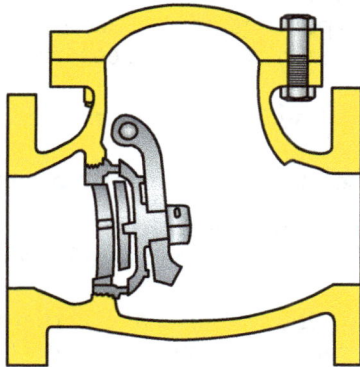

(B) SWING-CHECK VALVE

Figure 2 Check valves.

Figure Credit: Image courtesy of Watts Regulator Company

Figure 3 Relief valves.

Additional Resources

Pipe, Fittings, Valves, Supports, and Fasteners, International Pipe Trades Joint Training Committee. 2000. Annapolis, MD: United Association.

ASTM F683-10, Standard Practice for Selection and Application of Thermal Insulation for Piping and Machinery. ASTM International. Available at **www.astm.org/DATABASE.CART/HISTORICAL/F683-10.htm**

1.0.0 Section Review

1. A balancing valve is used to _____.
   a. mix two separate fluid flows
   b. divert one fluid flow into two
   c. maintain a specific amount of flow
   d. start and stop flow

2. Non-flanged valves tend to be used on _____.
   a. cold pipe applications
   b. hot pipe applications
   c. smaller pipe sizes
   d. pipe systems carrying steam

3. Insulation on a flanged valve should extend to what distance from the flange bolt heads?
   a. Twice the insulation thickness
   b. Equal to the insulation thickness
   c. Half the insulation thickness
   d. The diameter of the flange

2.0.0 FITTINGS AND JOINTS

Objective

Describe the different types of fittings and joints and their insulation requirements.

a. Describe the different types of fittings and joints.
b. Explain the insulation practices related to non-flanged fittings and joints.
c. Explain the insulation practices related to flange fittings.
d. Explain the insulation practices related to mechanical fittings.

Performance Tasks

3. Apply insulation to flanged fittings.
4. Apply insulation to non-flanged fittings.
5. Apply insulation to mechanical fittings.

Trade Terms

Heel: The inside curve of an elbow.

Hydraulic setting insulation cement: A type of quick-setting cement.

Long-radius elbow: A 45-degree or 90-degree elbow with its radius equal to one and a half times the diameter of the line.

Miters: Segments of insulation that fit together to insulate an elbow.

Radius: The distance between the center point and the perimeter of a circle or arc.

Short-radius elbow: A 45-degree or 90-degree elbow with its radius equal to the diameter of the line.

Throat: The outside curve of an elbow.

Fittings allow a pipe system to change direction, pipe size, or even the type of pipe. Fittings are also used to connect a pipe system to equipment, such as pumps or tanks.

2.1.0 Pipe Fitting Types

Pipe fittings are used to make pipe connections. These connections can be between two or more sections of the same type and size of pipe, between different sizes of pipe, or to terminate a pipe section. Fittings can be the following:

- *45-degree ells* – Used to change pipe direction by 45 degrees.
- *90-degree ells* – Used to change pipe direction by 90 degrees.
- *Return bends* – Used to change pipe direction by 180 degrees, i.e., toward the direction from which the pipe came. Sometimes called a *U-bend*.
- *Tees* – A typical tee pipe fitting has three openings and is generally used to provide a single branch line from a main run of pipe. They can also be used to make two branches, each at 90 degrees from the main pipe. However, the latter is not generally allowed in piping systems with a significant amount of flow. A 4-way tee has four openings and is used to make three branches, each at 90 degrees from the main pipe.
- *Reducers* – Used to connect two different pipe sizes.
- *Caps* – Used to terminate a run of pipe.

Ells are sometimes referred to as *elbows*, although the term elbow tends to be used for 90-degree ells only.

90-degree elbows can have a turning radius that is either short or long (*Figure 4*). The turning radius of a short-radius elbow is equal to the diameter of the pipe. The turning radius of a long-radius elbow is equal to one and one-half times the diameter of the pipe. (In *Figure 4*, X represents the diameter of the pipe.) Most welded steel pipe elbows requiring insulation are long-radius, unless space does not allow for a long-radius elbow.

Some fittings are the same outside diameter as the pipe in which they are used. Other fittings are considerably larger than their pipe and require either a larger insulation size or custom-fabrication insulation.

2.1.1 Fitting Attachment

Fittings can be connected to a pipe system in several ways. Some pipe systems use more than one type of connection. Connection methods include the following:

- *Threaded joints* – Attachment by means of male and female threads.

Did You Know?

The term *pipe fitting* is used for several types of pipe connections. A craftworker whose trade is to assemble pipe systems is called a *pipefitter*.

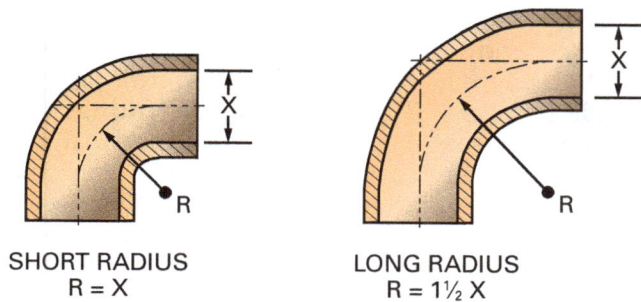

Figure 4 Short- and long-radius elbows.

- *Soldered and brazed joints* – Attachment by means of heating and melting a filler metal, which flows into the joint and seals it. Soldering takes place at lower temperatures than brazing, but neither process melts the base metal (as welding does). Also called *sweated joints*.
- *Welded joints* – Connections made by melting two surfaces together. In most metal welding processes, a filler metal is added to fill the joint.
- *Cemented (glued) joints* – Plastic fittings, such as PVC, are cemented together. One solution is applied to the parts to break down the glossy finish and clean the joint surfaces. Then the cement is applied and a period of time is required for the joint to cure.
- *Mechanical couplings* – Connections held together by clamps, rings-and-grooves, or other mechanical means.

2.1.2 Fitting Materials for Specific Pipe Systems

Copper fittings (*Figure 5*) are only used on copper pipe. Copper fittings, like copper pipe, have comparatively thin walls, and are approximately the same outside diameter as the pipe. Copper pipe

sections are joined by soldered or brazed couplings. Brazing is more common in industrial applications where pressures and temperatures are higher. The following facts apply to copper pipe systems:

- Large copper fittings are occasionally flanged.
- Copper elbows can be long radius or short radius. Short-radius elbows are used where pressure loss from friction in the pipe is not of great concern. Refrigerant piping for example, which often must be insulated, should use long-radius elbows exclusively, since short-radius elbows create more pressure drop.
- Valves used in copper piping are typically made from brass, since copper is generally too soft for valve construction. Many brass valves are threaded.

Steel pipe fittings (*Figure 6*) are usually made of black steel or stainless steel. Steel pipe fittings are joined with threaded or welded joints. The following facts apply to steel pipe systems:

- Threaded joints are common on most pipe sizes up to 2" (DN50).
- Welded joints are more common with pipe sizes over 2" (DN50).
- Grooved joint and clamp connectors (mechanical couplings) are used only on low-pressure water pipe systems.
- Most threaded 90-degree elbows are short radius.
- Most welded elbows for pipe sizes up to 2" (DN50) are short radius.
- Elbows are typically long radius, except for threaded-joint elbows over 2", which are commonly short radius.

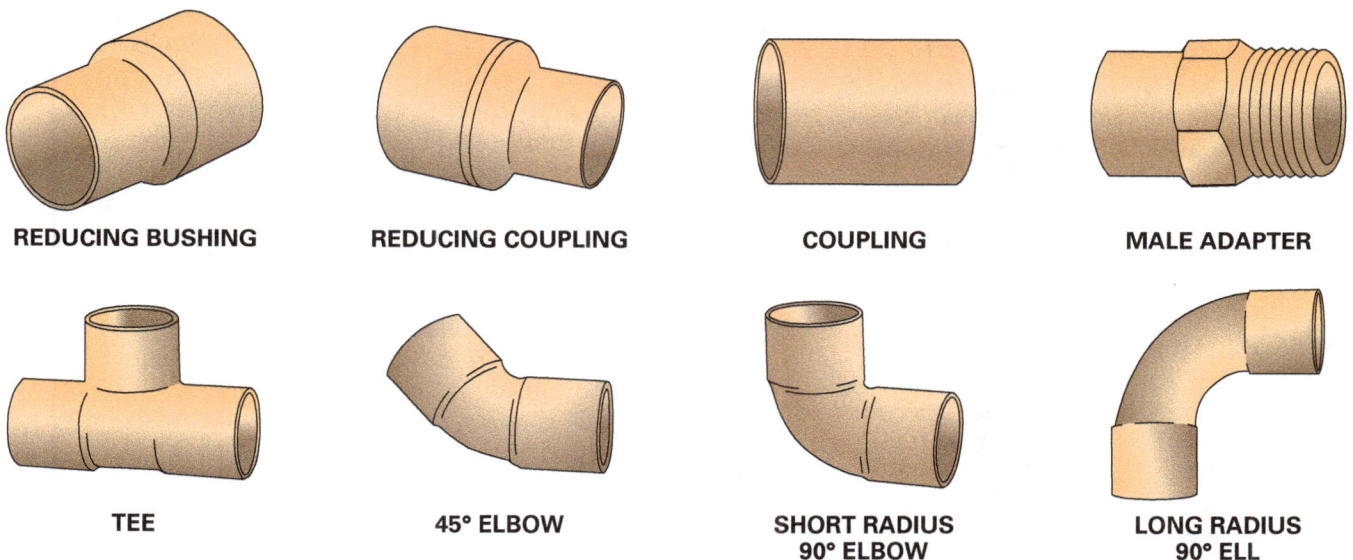

Figure 5 Copper fittings.

90-DEGREE ELL

45-DEGREE ELL

Figure 6 Steel pipe fittings.

The configuration of plastic fittings (*Figure 7*) is similar to that of steel fittings. Valves in plastic piping systems are often brass or steel, but PVC valves are also used in some applications. Brass and steel valves are attached to plastic pipe with a threaded male or female adapter. The following facts and guidelines apply to plastic pipe systems:

- Threaded brass valves are commonly used on pipe up to 2" (DN50). Flanged valves are generally used on larger sizes.
- Threaded fittings are larger in outside diameter than the attached pipe to accommodate the threads and maintain a reasonable wall thickness. If the insulation is used on pipe ½" or less (≤DN15), it can fit over the fitting with no loss of performance. For pipe sizes greater than ½" (DN15), larger insulation will be called for.

Cast iron fittings are used on drainage pipe systems and similar low-pressure applications. No-hub fittings are attached to cast iron pipe by banded clamps (*Figure 8*). Hubbed fittings are designed so that the end of the pipe inserts into a belled hub. Due to the thickness of cast iron pipe and the resulting diameter of the hub, it is necessary to insulate the hub section with an oversized section of insulation. However, cast iron pipe and fittings are rarely used in applications that require it to be insulated.

2.2.0 Insulating Pipe Fittings

Fitting insulation varies according to the type of pipe system being insulated. As a rule of thumb, the guidelines for insulating the pipe system will be applicable to insulating the pipe system fittings.

There are several strategies for insulating fittings. Fitting insulation can fall into any of the following categories:

- Factory-molded to fit a specific fitting
- Machined from insulation materials
- Fabricated from pipe insulation, blanket, or other suitable insulation material

Fabrication of fittings can be performed on site or in an insulation shop.

2.2.1 PVC Fitting Covers

PVC fitting covers (*Figure 9*) are widely used in commercial and industrial applications. PVC covers make pipe fittings look better and are economical, resistant to weather and many chemicals, and relatively easy to install. They are used to cover and protect fiberglass insulation and similar products from the elements, and to provide an appealing, cleanable finish to the installation.

PVC covers are widely used on elbows, tees, reducers, end caps, and valves. The covers are available for different pipe sizes and thickness combinations. Fitting covers are usually delivered to the job site in cartons and are marked with manufacturer-specific identifying numbers (*Table 1*). Each box contains both the PVC covers and the fiberglass inserts.

With a little fabrication, PVC covers can be used over a wide variety of fittings other than those they were designed for. For example, ells can be insulated with pre-fabricated covers, then finished with a PVC cover. Valves can be insulated with pipe insulation sized to the outside diameter of the valve flange, while the bonnet can be insulated with blanket insulation. The entire unit can then be covered with PVC.

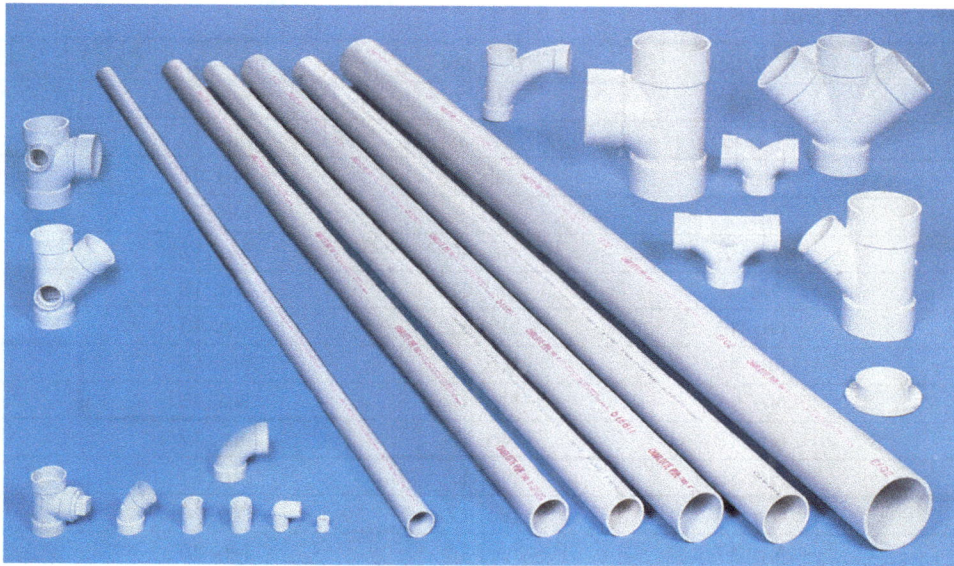

Figure 7 Plastic fittings.

BANDED CLAMPS

Figure 8 No-hub cast iron pipe and fittings.

Covers can be installed with most pipe insulations, but there are limitations. PVC will soften and begin to deform at approximately 150°F (66°C). Consequently, insulation installed under the PVC should allow the surface of the insulation to reach no more than 125°F (52°C). If the insulation surface will approach 125°F (52°C), the cover should be applied loosely.

Some PVC covers are susceptible to damage from ultraviolet light. If using a cover that can be damaged by sunlight, do not use that cover for outdoor applications.

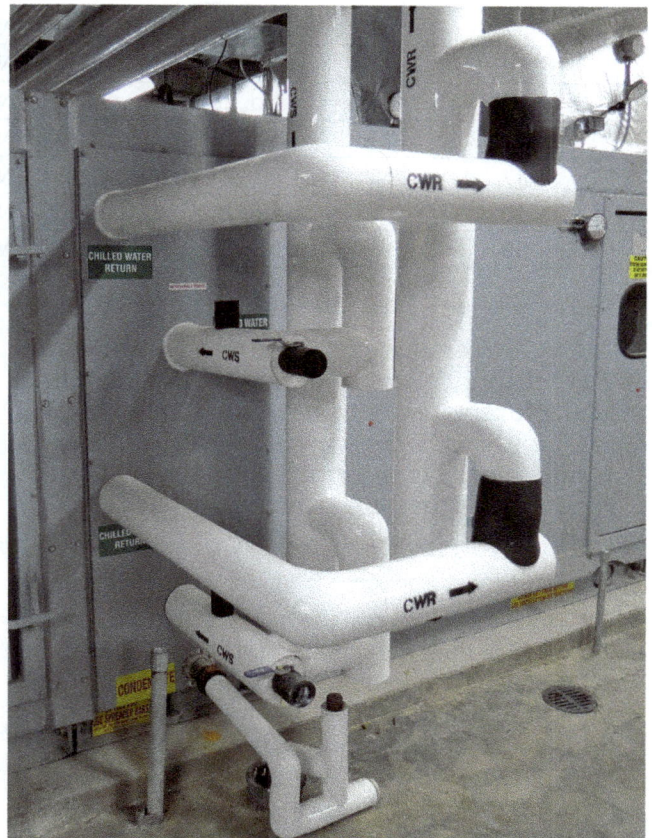

Figure 9 PVC-covered pipe and fittings.

Table 1 PVC Covers for IPS and CTS Pipe

Iron Pipe Size	½" Thick	1" Thick	1½" Thick	2" Thick
½"	No. 3	No. 7	No. 10	No. 12
¾"	No. 5	No. 7	No. 10	No. 12
1"	No. 5	No. 9	No. 11	No. 13
1¼"	No. 7	No. 9	No. 12	No. 13
1½"	No. 7	No. 10	No. 12	No. 13
2"	No. 9	No. 11	No. 13	No. 15
2½"	No. 10	No. 12	No. 13	No. 15
3"	No. 11	No. 13	No. 15	No. 17
3½"	No. 12	No. 15	No. 15	No. 17
4	No. 13	No. 15	No. 17	No. 18
4½"	No. 15	No. 17	No. 17	No. 18
5	No. 16	No. 17	No. 18	No. 19
6	No. 17	No. 18	No. 19	No. 20
8	No. 19	No. 20	No. 21	No. 22
10	No. 21	No. 22	No. 23	No. 24
12	No. 23	No. 23	No. 25	No. 26
14	No. 24	No. 24	No. 26	No. 27
15	No. 25	No. 25	No. 27	No. 28
16	No. 26	No. 26	No. 28	No. 29
17	No. 27	No. 27	No. 29	No. 30
18	No. 28	No. 29	No. 30	—
19	No. 29	No. 30	—	—
20	No. 30	—	—	—

A PVC cover size chart for one manufacturer of PVC fitting covers is shown in *Table 1*. (Other manufacturers use similar charts.) To use the chart, cross-index the pipe size with the insulation thickness. The resulting number is the identifier of the fitting cover sized for that pipe in insulation thickness.

> **NOTE**
>
> The numbers above the heavy line in *Table 1* will fit both long-radius and short-radius elbows. Long-radius elbow covers for sizes below the heavy line will come in cartons marked for pipe size and insulation thickness (for example, 6 × 1½ L.R. [*long radius*]). The exact sizes that fit both long and short radius ells will vary slightly according to manufacturer.

If pipe insulation is ½" thick and elbows are oversized, regular PVC covers can be used by installing an extra layer of pipe insulation adjacent to the fitting. For example, when insulating a 4" pipe with ½" thick insulation, installing an additional layer of 5 × ½ insulation adjacent to the fitting will allow the insulation to build up over the fitting. The fitting size for the cover would then be 4 × 1.

Follow these guidelines for a professional PVC cover installation:

- The pipe insulation must be installed on each side of the fitting to provide a firm base for the fitting cover.
- After identifying the cover number, select the correct number of covers for the task from storage.
- If the covers are being installed on a pipe system that is operating below ambient temperatures, then the joints should be sealed with a vapor retardant mastic during installation. Fiberglass butt strips and PVC tape are sometimes used as well, but it is difficult to achieve a good seal with tape alone. Review the job specifications to verify the need for vapor retardant, as well as the required type.
- If the covers are being installed on a hot pipe system, fiberglass inserts must be installed to reduce the heat at the PVC cover.
- If the cover is close to a fitting, or if the installation is in a tight area, the cover may require

trimming for a good fit. Use scissors for a clean trim.

- If the pipe finish will be ASJ, apply the butt strip of the insulation over the PVC cover for a neat application. This will also act as a vapor retarder.
- PVC covers installed outside, or in areas subject to water splash, spray, or runoff, should be sealed similarly to vapor sealing.
- Cold pipe systems can be insulated with PVC covers filled with foam-in-place insulation.

> **WARNING!**
>
> Be sure to use the correct attachment method for the type of insulation so that the insulation is not damaged.

2.2.2 Molded or Machined Insulation Fittings

Molded or machined insulation fittings are available in all common pipe insulation materials.

> **NOTE**
>
> The term *insulation fitting* is used to describe a molded or machined piece of insulation that is used to cover a pipe system fitting. A pipe fitting itself is a piece of hardware, such as a tee, an ell, or a reducer.

The advantages of molded or machined insulation fittings are as follows:

- There is an exact fit of the insulation to the pipe fitting.
- The material is identical to the pipe insulation, with the same characteristics.

Molded or machined insulation fittings are available for tees, reducers, and other fittings but are most commonly used on ells.

Pre-formed insulation fittings can be attached by wire, tape, or adhesive. If the insulation fittings are cellular glass or similar closed-cell foam material, the joints should be sealed with vapor retardant mastic.

Pre-formed insulation fittings can be vapor sealed and finished to match the pipe insulation. If the insulation fittings are fiberglass, they can be finished with mastic and reinforcing fabric, PVC covers, or left unfinished on concealed hot pipe applications. The insulation fittings may also be finished with aluminum or other materials.

> **CAUTION**
>
> If the pipe system is cold, the insulation fittings must be vapor sealed. If vapor retardant mastic is used, the entire fitting must be coated and sealed over to the vapor barrier on the pipe insulation.

2.2.3 Mitered Segment Insulation Fittings

Elbows can be insulated by cutting the pipe insulation into segments that fit around the elbow. These segments are called miters.

Because the inside measurement of the elbow is much shorter than the outside measurement, miters are shaped like trapezoids (*Figure 10*). The wide end is referred to as the heel, and the narrow end is referred to as the throat.

> **NOTE**
>
> If you are cutting miters from direct measurements of the elbow, remember that you are measuring distance to be covered by the inner diameter of the insulation. The cuts have to be laid out on the inside of the insulation. If they are laid out on the outside of the insulation, the insulation will not fit properly.

With variations in measurements and cuts, miters can be made to fit any size elbow. *Table 2* and *Table 3* define the heel and throat measurements (B and C), and the number of miters required (A), for various sizes of long radius and short-radius elbows. *Figure 11* shows how miters fit together when cut.

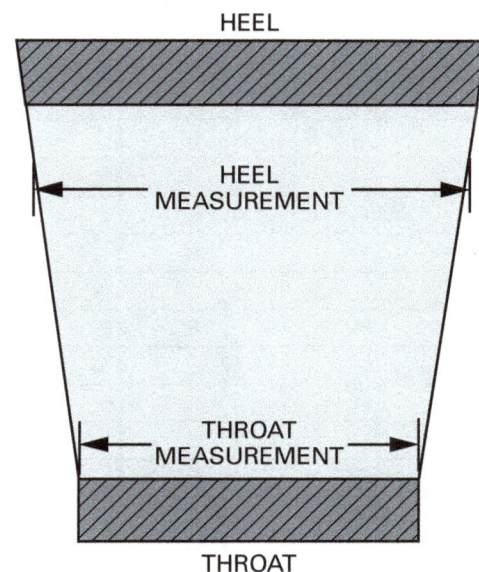

Figure 10 Miter shape and measurement locations.

Table 2 Miter Dimensions and Quantity for Long-Radius Elbows

Pipe Size	1" Thick			1½" Thick			2" Thick			2½" Thick			3" Thick		
	A	B	C	A	B	C	A	B	C	A	B	C	A	B	C
2½	4	2½	½	4	2¾	¼	—	—	—	—	—	—	—	—	—
3	4	2⅞	¾	4	3	½	3	4⅜	⅜	3	4⅝	⅛	—	—	—
3½	4	3⅛	1	4	3⅜	¾	4	3½	⅝	4	3¾	⅜	3	3¼	¼
4	6	2⅜	¾	6	2½	⅝	6	2¾	½	6	2⅞	⅜	4	4½	¼
4½	6	2⅝	⅞	6	2¾	¾	6	2⅞	⅝	6	3	½	6	3⅛	⅜
5	6	3	1	6	3⅛	⅞	6	3¼	¾	6	3⅜	⅝	6	3½	⅜
6	6	3½	1½	6	3⅝	1⅛	6	3¼	1	6	3⅞	⅞	6	4	¾
7	—	—	—	6	4⅛	1⅜	6	4¼	1¼	6	4⅜	1⅛	6	4½	⅞
8	—	—	—	8	3½	1¼	8	3⅝	1⅛	8	3¾	1	8	3¾	⅞
9	—	—	—	8	3⅞	1⅜	8	4	1¼	8	4⅛	1¼	8	4¼	1⅛
10	—	—	—	8	4¼	1⅝	8	4⅜	1½	8	4½	1⅛	8	4⅝	1¼
11	—	—	—	8	4¾	1¾	8	4¾	1⅝	8	4⅞	1⅝	8	5	1½
12	—	—	—	8	5⅛	2	8	5⅛	1⅞	8	5¼	1¾	8	5¾	1⅝
14	—	—	—	8	5¾	2½	8	5⅞	2⅜	8	6	2¼	8	6⅛	2⅛
16	—	—	—	8	6½	2⅞	8	6⅝	2¾	8	6¾	2⅝	8	6⅞	2⅝
18	—	—	—	8	7⅜	3¼	8	7⅜	3⅛	8	7½	3	8	7⅝	3
20	—	—	—	8	8⅛	3⅝	8	8¼	3½	8	8¾	3½	8	8⅜	3⅝
22	—	—	—	10	7¾	3½	10	7⅞	3½	10	7⅞	3⅜	10	8	3¼

Table 3 Miter Dimensions and Quantity for Short-Radius Elbows

Pipe Size	1" Thick			1½" Thick			2" Thick			2½" Thick			3" Thick		
	A	B	C	A	B	C	A	B	C	A	B	C	A	B	C
3½	4	2½	¼	2	5⅜	⅛	—	—	—	—	—	—	—	—	—
4	3	3¾	¼	2	6⅛	⅛	—	—	—	—	—	—	—	—	—
4½	4	3	½	4	3¼	¼	2	6⅞	⅛	—	—	—	—	—	—
5	4	3½	½	4	3⅝	¼	2	7⅝	⅛	—	—	—	—	—	—
6	6	2⅝	½	6	2¾	⅜	4	4⅜	¼	2	9¼	⅛	—	—	—
7	—	—	—	6	3¼	½	6	3⅜	¼	4	3¼	¼	—	—	—
8	—	—	—	6	3⅝	⅝	6	3¾	½	6	3⅞	¼	3	8⅛	¼
9	—	—	—	8	3	½	8	3⅛	⅜	8	3¼	¼	6	4⅜	¼
10	—	—	—	8	3⅜	⅝	8	3⅜	½	8	3½	⅜	8	3⅝	¼
11	—	—	—	8	3⅝	¾	8	3¾	⅝	8	3¾	½	8	3⅞	⅜
12	—	—	—	8	3⅞	¾	8	4	¾	8	4⅛	⅝	8	3⅝	¼
14	—	—	—	8	4⅜	1⅛	8	4½	1	8	4⅝	⅞	8	4¾	¾
16	—	—	—	8	5	1¼	8	5⅛	1¼	8	5⅛	1⅛	8	5¼	1
18	—	—	—	8	5⅝	1½	8	5⅝	1⅜	8	5¾	1¼	8	5⅞	1¼
20	—	—	—	8	6⅛	1⅝	8	6¼	1⅝	8	6⅜	1½	8	6½	1⅜
24	—	—	—	8	7⅜	2⅛	8	7⅛	2	8	7½	1⅞	8	7⅝	1¾

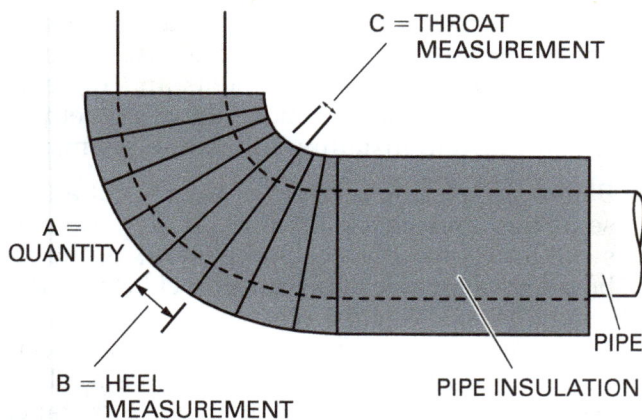

Figure 11 Miter fit.

If the elbow is not available for direct measurement, first determine the radius. The radius of a long-radius elbow is equal to the nominal pipe size multiplied by 1.5. The radius of a short-radius elbow is equal to the nominal pipe size.

Use *Table 2* or *Table 3* as appropriate to determine the heel and throat lengths and the number of miters needed.

There are two methods of calculating heel and throat measurements and the number of required miters. The first method is simpler, but the second method is more precise.

For the simplest method, use the following steps:

Step 1 Measure the heel and throat of the elbow using a tape measure.

Step 2 Use *Table 4* or *Table 5* with the pipe size to determine the required number of miters. (If the elbow is long-radius, use *Table 4*. If it is short-radius, use *Table 5*.)

Step 3 Divide the heel measurement of the elbow by the required number of miters to determine the heel measurement for each miter.

Step 4 Divide the throat measurement of the elbow by the required number of miters to determine the throat measurement for each miter.

For example, if you are insulating a 6" short-radius elbow, *Table 5* tells you that you will need 6 miters. If the heel measurement of the elbow is 20", then the heel measurement for each miter will be $3\frac{1}{3}$" (20" ÷ 6 miters = $3\frac{1}{3}$"). If the throat measurement of the elbow is 6", then the throat measurement for each miter will be 1" (6" ÷ 6 miters = 1").

Table 4 Number of Miters by Pipe Size (Long-Radius Elbow)

Pipe Size	Number of Miters Required
≤ 4"	4
5" to 7"	6
8" to 10"	8
11" to 18"	10
20" to 30"	12

Table 5 Number of Miters by Pipe Size (Short-Radius Elbow)

Pipe Size	Number of Miters Required
≤ 2"	3
3" to 4"	4
5" to 7"	6
8" to 10"	8
11" to 18"	10
20" to 30"	12

To calculate the miter dimensions with the more precise method, use the following steps:

Step 1 Calculate the radius of the elbow. If it is a long-radius elbow, do this by multiplying the pipe diameter by 1.5. If it is a short-radius elbow, the radius is equal to the pipe diameter (pipe diameter × 1).

Step 2 Use the radius with the following formula to find the heel measurement of each miter:

$$\frac{[(2 \times \text{radius}) + \text{OD of insulation}] \times \pi}{\text{number of miters} \times 4}$$

Step 3 Use the radius with the following formula to find the throat measurement of each miter:

$$\frac{[(2 \times \text{radius}) - \text{OD of insulation}] \times \pi}{\text{number of miters} \times 4}$$

For example, to determine the heel measurement of a 6" long-radius 90-degree elbow, use the following steps:

Step 1 Calculate the radius of the elbow by multiplying the nominal pipe diameter by 1.5:

Radius = nominal pipe size × 1.5
Radius = 6" × 1.5
Radius = 9"

Step 2 In this example, the insulation has an outside diameter (OD) of 10.5". Use this measurement and the radius with the appropriate formula to find the heel measurement of each miter:

$$\frac{[(2 \times radius) + OD\ of\ insulation] \times \pi}{number\ of\ miters \times 4} =$$

$$\frac{[(2 \times 9") + 10.5"] \times 3.14}{6\ miters \times 4} =$$

$$\frac{[18 + 10.5"] \times 3.14}{24} =$$

$$\frac{28.5" \times 3.14}{24} =$$

$$\frac{89.49"}{24} = 3.73"\ (approx.\ 3\frac{3}{4}")$$

Step 3 Use the other formula to find the throat measurement of each miter:

$$\frac{[(2 \times radius) - OD\ of\ insulation] \times \pi}{number\ of\ miters \times 4} =$$

$$\frac{[(2 \times 9") - 10.5"] \times 3.14}{6\ miters \times 4} =$$

$$\frac{[18 - 10.5"] \times 3.14}{24} =$$

$$\frac{7.5" \times 3.14}{24} =$$

$$\frac{23.55"}{24} = 0.98"\ (approx.\ 1")$$

2.2.4 Cutting and Installing Miters

Miters can be cut by hand, or with a bandsaw. If using a bandsaw, it must be a unit designed to cut insulation. To use a bandsaw to cut miters, set the angle bar to the number of miters required, and set the back-stop equal to the heel measurement.

> **WARNING!**
>
> When using a bandsaw, or any power construction equipment, be sure to follow all safety precautions and wear suitable eye protection. Be sure that all power machinery is properly guarded according to OSHA standards.

If cutting miters by hand, remember to make the first heel and throat measurements from as close to the end of the insulation as possible (*Figure 12*). In order to minimize waste, start the first cut no more than a quarter of an inch from the edge of the insulation.

Whether cutting miters by hand or with a bandsaw, be sure to determine and use the correct heel and throat measurements and the correct number of miters. Follow these guidelines when cutting and installing mitered segments:

- Before completing a group of miters, cut a few segments and check their fit. If they do not seem to fit properly, check the radius of the elbow and the values on the miter charts, or the results of your calculations.
- Miter halves on either side of the pipe can be cemented together, or they can be applied without cement and held with wire or tape. Either way, the miters should fit without voids. Miter pairs can be tweaked slightly for fit.
- Pipe insulation should be complete to the center of the weld joint before miters are installed. The first set of miters should be fitted snugly against the pipe insulation.
- If necessary for a good fit, trim the pipe insulation where it contacts the weld joint. This may be a particular concern when using calcium silicate.
- The last pair of miters will complete the installation of the pipe insulation on the opposite side. The miters and the pipe insulation may need to be trimmed at the back (heel) of the elbow, especially if the insulation is thick.
- If the application is a cold pipe system, all miters must be sealed. If using closed-cell material, the vapor seal can be accomplished by applying a vapor retardant mastic to each joint as it is applied. If the insulation is fiberglass, it can be sealed by fitting with aluminum foil and taping the foil closed with cloth tape. Alternatively, the fittings can be finished with mastic and reinforcing fabric, or with a PVC cover.

2.2.5 Single Miter Pipe Section (Stove Pipe) Fitting Covers

Mitered insulation for smaller pipes can be fit with a single cut (*Figure 13*). Make the cut equal to one-half of the elbow bend in degrees. For a 45-degree bend, miter each side to $22\frac{1}{2}$ degrees. For a 90-degree elbow, miter each side to 45 degrees. Stove pipe covers are widely used with flexible foam insulation.

Stove pipe miters can also be made for screw-thread fittings by using an insulation size that sleeves over the pipe insulation. As always, cold pipe applications must be sealed with adhesive or mastic.

Follow these guidelines for mitering stove pipe fitting covers:

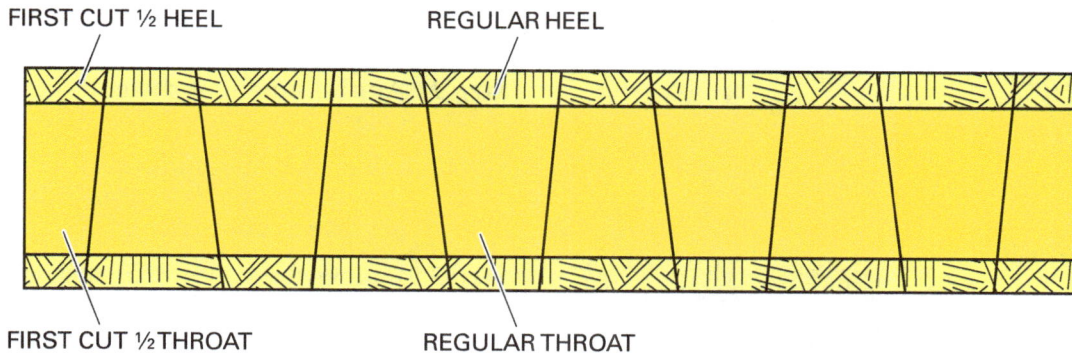

FIRST CUT ½ HEEL   REGULAR HEEL

FIRST CUT ½ THROAT   REGULAR THROAT

Figure 12 Half-measurement on initial miter cut.

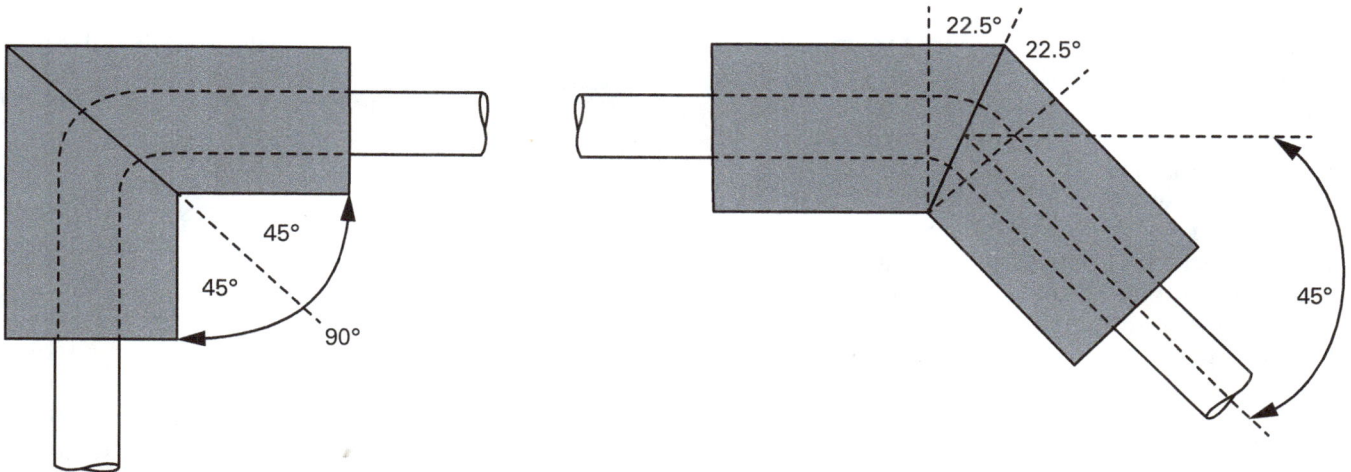

Figure 13 Single-miter insulated ells.

- If the application is on copper pipe, the fitting can be cut in the regular pipe insulation. There is no need to stop the insulation at the joint.
- For more precise and better-fitting miters, use a miter box instead of freehanding the miter cuts.

2.2.6 Three-Piece Miter Pipe Section

Mitered insulation for piping that is 1½" in diameter and larger typically cannot be made using a single cut. One reason is that the 90-degree elbows used with large diameter piping have extended throat lengths. Because of these size concerns, it is best to insulate these 90-degree elbows using the three-piece miter method.

The following steps should be used when making three-piece miters:

Step 1 Cut the first piece of insulation so that one end is at a 22½-degree angle and the other end is a 90-degree straight cut. The throat should be approximately ½". (The throat measurement will get longer as the pipe size gets larger.)

Step 2 Cut the second piece (the center piece) of insulation so that both ends are at a 22½-degree angle. As with the first cut, the throat should be approximately ½".

Step 3 Cut the third and final piece the same way as the first piece—with a 22½-degree angle on one end and a 90-degree straight cut at the other end.

Step 4 Brush all of the 22½-degree angle ends with a thin coat of adhesive, and press them together after the adhesive gets tacky.

Step 5 Once the miters have been glued together, cut open the throat of the section and apply a thin coat of adhesive to both sides.

Step 6 Place the open three-piece miter over the pipe fitting and press it together to secure it.

2.2.7 Built-Up Insulation for Fittings

Pipe fittings can be insulated by building up layers of material such as insulation cement or blanket insulation. Built-up covers of insulation cement should be installed only on hot pipe systems. Built-up blanket insulation can be installed on both hot and cold pipe systems.

There are three types of cements used for insulating fittings:

- Hydraulic setting insulation cement
- High-temperature cement
- Finish cement

Hydraulic setting insulation cement (also known as *one-coat cement*) is made of mineral wool fibers, clays, binders, and portland cement. Mix only the amount of cement that can be used in two hours. This cement sets and cannot be remixed. Hydraulic setting insulation cement is used on pipe that will operate up to temperatures of 500°F (260°C), and is used both as an insulating and finishing cement. Hydraulic setting insulation cement can also be used to make a smooth finish over insulated fittings.

High-temperature cement is similar to hydraulic setting insulation cement except it does not contain portland cement and has a higher mineral wool content. High-temperature cement does not chemically set and must dry to become rigid. It shrinks considerably as it dries. If appearance is a factor, the job may require a second coat.

Finish cements are made from gypsum or other material that make a smooth uniform finish. These cements are applied over other cements after they have been applied and allowed to dry. This cement may be used to make a smooth finish over mitered segments of calcium silicate.

Insulation cement fittings are most commonly used on small pipe fittings when the pipe insulation being installed is calcium silicate. The cement is packaged in 50-pound bags that should be stored in a dry place. When a small number of fittings are to be insulated, fill a pail to approximately ¼ with clean water, add cement and mix thoroughly. Use rubber gloves when mixing and using cement. The cement should be mixed to a stiff consistency. If the mixture is dry or extremely stiff, add water sparingly.

> **WARNING!**
> As with any industrial material, read and understand the safety data sheet (SDS) for cement before use. Cements can cause burns, and if inhaled they can burn the lungs and sinuses. In powdered form they can cause damage to the eyes as well. Review the SDS for any cements before using them. Follow all applicable safety guidelines. An example of an SDS is shown in *Figure 14*.

Follow these guidelines to create a built-up fitting cover:

Step 1 Apply cement at the ends of the pipe insulation first.

Step 2 Press the cement firmly in place. Use a trowel, or apply by hand (using rubber gloves for protection).

Step 3 Follow the contour of the fitting. Apply cement to a thickness equal to or slightly greater than the pipe insulation.

> **NOTE**
> Small fitting covers can be formed with one coat of cement. However, it may be easier to apply a light initial coat, wait for it to set, then install a second finishing coat.

Step 4 Complete the top of the fitting before moving to the bottom. This lets gravity work with you as you press cement from the opposite side, helping keep the cement in place.

Step 5 If a heated system must be in use while you are applying cement, wrap the metal surfaces of the fitting with canvas or reinforcing fabric dipped in lagging adhesive.

Step 6 If the fitting cover is exposed, finish it with lagging adhesive, PVC covers, aluminum covers, or mastic per job requirements. (Built-up cement fitting covers may be left unfinished if they are concealed.)

2.2.8 Flexible Insulating Blanket

An insulating blanket can be cut into strips and used to insulate a fitting. The strips are wrapped snugly around the fitting. As far as possible, no voids are left. Follow these guidelines for a professional installation:

Step 1 Tuck ends and edges of the blanket in to make the shape as uniform as possible.

Step 2 Wrap the blanket with tape or twine to hold it in place. If a final finish is required, it can be applied after the blanket is secured.

Step 3 If the fittings are part of a cold pipe system, vapor seal them.

Trade Names:

13/16" Micro-Aire® Duct Board	Micro-Lok® HP
800 Series Spin-Glas® Board Insulations, faced	Micro-Lok® HP Ultra
Grooved Duct Board	Micro-Lok® Pipe Insulation
Mat-Faced Micro-Aire® Duct Board	R Series Microlite® (FSK, PSK, & vinyl faced)
Micro-Flex™ Large Diameter Pipe and Tank Wrap	Spiracoustic Plus™
Microlite® Standard	SuperDuct™ RC Boards
Micro-Lok® HP	

Section 2 - Hazards Identification

Emergency Overview

Inhalation of excessive amounts of dust from the product may cause temporary upper respiratory irritation and/or congestion-- remove individual to fresh air.

In high temperature applications, treatment, curing, or in geographic areas of high heat and humidity, this product may release gases irritating to the eyes, nose and throat.

Inhalation

Temporary mechanical irritation may occur upon exposure to dust or fibers released from cutting this product.

Irritation of the upper respiratory tract, coughing, and congestion may occur in extreme exposures. Severe irritation of the mouth, nose, and throat, as well as signs of central nervous system depression (drowsiness, dizziness, headache), may occur upon inhalation of vapors or gases.

Skin

Temporary irritation (itching) or redness may occur.

Figure 14 Example safety data sheet.

2.3.0 Insulating Line Flanges

Line flanges are used to connect one section of pipe to a pipe fitting or other piece of equipment (*Figure 15*). Flanges can be a welded to pipe or joined to it by threading. A pair of flanges can be used to join lengths of pipe. Once each flange is attached to a section of pipe, the two flanges are aligned and bolted together, typically with a gasket between the two flange faces.

Flanges can also be used to connect pipe to valves and other components. A single flange is attached to the pipe, and then is aligned with the flange of the device and bolted together.

Line flanges are insulated using a short section of larger pipe insulation to fit over the flange. Flange insulation should overlap the pipe insulation for a minimum distance equal to the thickness of the insulation.

Flange insulation covers may be fabricated in a shop using a bandsaw. The techniques of fabrication are similar to those for field fabrication. Molded flange covers are available in some insulation material. Prefabricated or molded covers are installed similarly to the field-fabricated covers previously described.

Table 6 lists the dimensions of insulation to fit standard pipe flanges. The insulation size indicates the standard pipe insulation size that will cover the flange. The thickness of the line flange insulation should be the same as the insulation used on the adjacent pipe. The chart shows lengths of insulation required to cover flanges using 1½", 2", and 3" thick insulation.

Overall, installing insulation on pipe flange connections is similar to insulating flanged valves. Follow these guidelines for a professional application.

- If the pipe insulation thickness is less than 2½ inches, install a short collar section (*Figure 16*) to each side of the flange so the OD of the insulation will equal the OD of the flange. Collar pieces can be cut from board or block insulation material.
- The pipe insulation should be stopped at a distance that allows the flange bolts to be removed. The end of the pipe insulation can be beveled for this purpose. If the application is on a cold system, the ends should be vapor sealed.

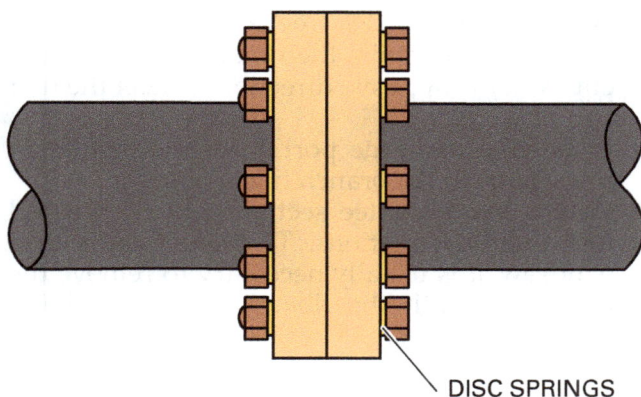

DISC SPRINGS

Figure 15 Assembled line flanges.

Table 6 Dimensions of Insulation for Line Flanges

Iron Pipe Size	OD of Flange	Insulation Size	"A" Thickness Flanges	"B" Inside Allowing for Bolts
½	3½	3	1½	2½
¾	3⅞	4	2	3
1	4¼	4	2¼	3¼
1¼	4⅝	5	2½	3½
1½	5	5	2¾	3¾
2	6	6	3	4
2½	7	7	3½	4½
3	7½	7	3¾	5¼
4	9	9	3¾	5¼
5	10	10	3¾	5¼
6	11	11	4	5½
8	13½	13	4½	6
10	16	16	4¾	6¼
12	19	19	5	6¾
14	21	21	5½	7
16	23½	24	5¾	7¼
18	25	25	6¼	7¾
20	27½	28	6¾	8¼
24	32	32	6¾	8¼

- After the flange cover is in place, check to make sure ends are flush and even with the collar pieces. If the collar pieces extend beyond the flange cover, trim flush with the cover. Check the flange cover for a snug fit. Fill any voids with scrap insulation. Check the flange cover and make sure it is level and plumb.

PVC covers are available to fit over the insulated pipe flanges.

2.4.0 Insulating Mechanical Fittings

Mechanical fittings can serve several purposes in a pipe system. They can allow the system to branch, reduce a pipe size along a run, or join two different types of pipe.

2.4.1 Insulating Tees

Tees can be insulated with the same material used on the pipe: molded or machined covers, or built-up blanket and cement. The basic application is the same as for elbows. Tee insulation should be at least as thick as the adjacent pipe insulation. Follow these guidelines for professional application of insulation at tees:

- Tee insulation can be cemented to the main insulation section, or installed separately.

- If on a cold pipe application, the insulation over the tee must be vapor sealed.
- When there are several tees, branches, or other obstructions, install the insulation for the first obstruction, then measure to the center of the next straight portion and cut the insulation section at the center point.

2.4.2 Saddle Cutting Insulation

Branch line insulation can be saddled into the mainline insulation. This is done by cutting a partial circle from the end of the branch line insulation to fit the outside diameter of the mainline insulation. Follow these guidelines for professional saddle cuts:

- When cutting circles freehand, make half the cut in one direction, then make the other half cut from the opposite direction to meet the first cut.
- Install the mainline portion of the insulation first, then cut the branch line to fit.
- With a bandsaw, tee sections can be mitered into mainline sections. To make cuts with a bandsaw, it is usually necessary to remove the jacket and split the fiberglass pipe covering into half sections.

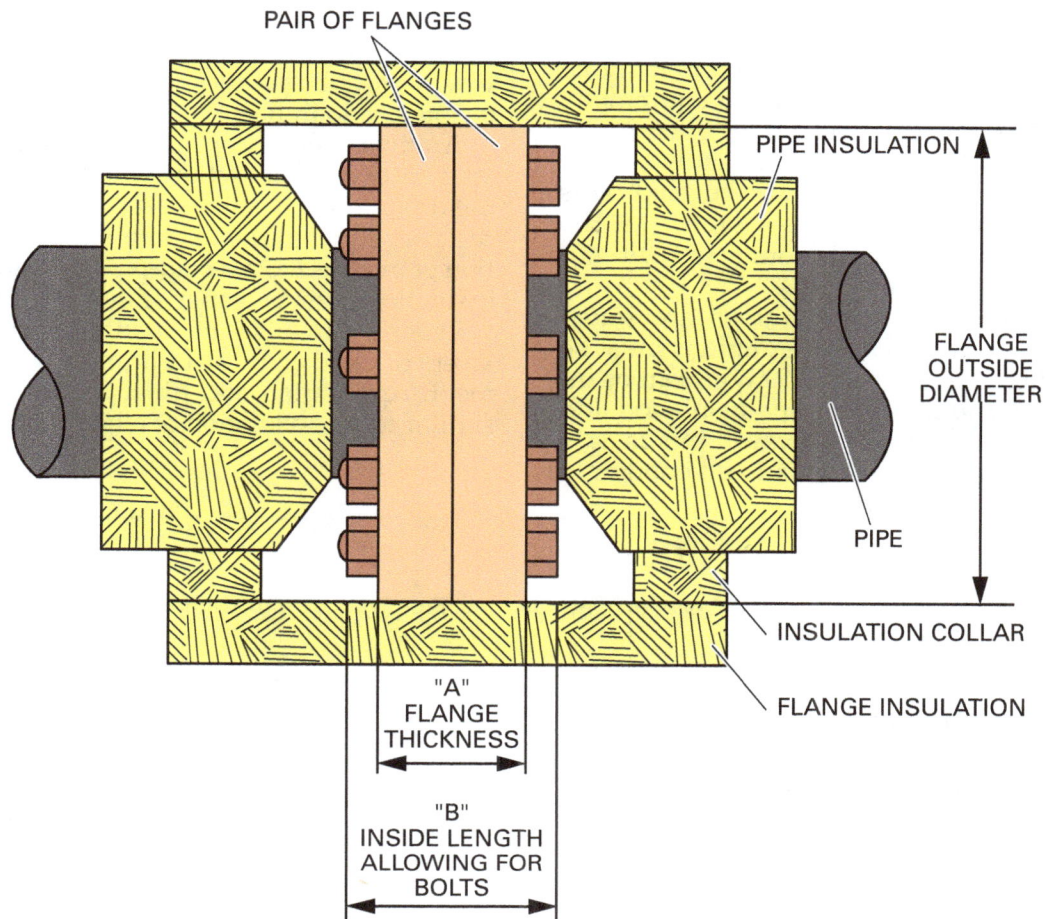

PAIR OF FLANGES

PIPE INSULATION

FLANGE OUTSIDE DIAMETER

PIPE

INSULATION COLLAR

FLANGE INSULATION

"A" FLANGE THICKNESS

"B" INSIDE LENGTH ALLOWING FOR BOLTS

Figure 16 Fabricated line flange cover.

2.4.3 *Insulating Reducer Covers*

Reducers are fittings that connect pipes of two different sizes. Reducers can be insulated in several ways. *Figure 17* shows two typical methods. The most common method is to run the larger size pipe insulation over the reducer. Some trimout will be required if the reducers are the screw or socket weld type. The insulation of the smaller pipe is butted up to the large pipe insulation. When using this method, be sure to fill the voids with a compatible insulation material.

Another common method is to stop the pipe insulation on both sides of the fitting. The fitting is then insulated with either built-up cement or layers of blanket. If the reducer is a cold-pipe fitting, the insulation must be vapor sealed.

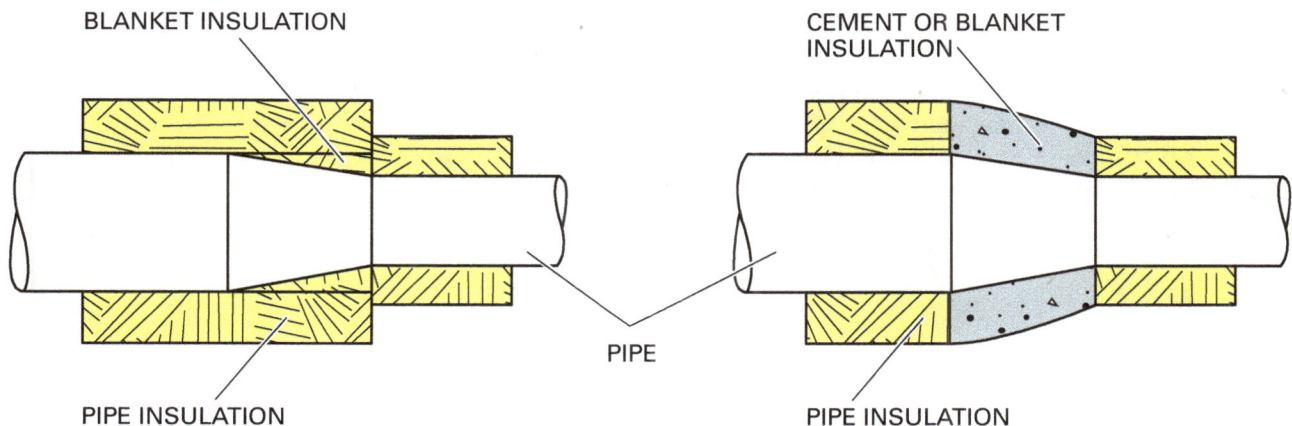

BLANKET INSULATION

PIPE INSULATION

PIPE

CEMENT OR BLANKET INSULATION

PIPE INSULATION

Figure 17 Reducer pipe insulation techniques.

2.4.4 Insulating Pipe Unions

Pipe unions are pipe connectors used to join two lengths of pipe or to connect pipe to equipment. They may have threaded, welded, or brazed connections. For threaded unions, the ends of pipe insulation at the unions should be beveled. This allows the placement of a pipe wrench at the union if the joint must be disconnected without damaging the insulation.

On hot pipe applications, unions are sometimes left uninsulated. However, unions must be insulated on cold pipe systems. Unions can be insulated by trimming out the pipe insulation and running it over the union. However, this is only possible if the pipe insulation is thick enough to allow the trim-out, keeping adequate thickness over the union. If the unions are insulated this way, some job specifications call for the finish to be marked to indicate the location of the union.

Another method of insulating unions is to install an oversized section of pipe insulation over the union. The oversized section should overlap a distance equal to the thickness of the pipe insulation being used. On a cold-pipe system, the union insulation must be vapor sealed.

Unions can also be insulated by wrapping layers of blanket, or by building a cover using cement. Application and finish in this method are similar to the insulation of elbows.

Additional Resources

Pipe, Fittings, Valves, Supports, and Fasteners, International Pipe Trades Joint Training Committee. 2000. Annapolis, MD: United Association.

2.0.0 Section Review

1. Copper pipe and fittings are typically joined by _____.

   a. cement
   b. hardware
   c. soldering or brazing
   d. threads or clamps

2. The narrow end of a miter is called the _____.

   a. throat
   b. heel
   c. snip
   d. bevel

3. Line flanges are insulated using _____.

   a. fiberglass mixed with resin
   b. scrap insulation
   c. insulating cement
   d. a section of regular pipe insulation

4. Which of the following statements about insulating pipe unions is true?

   a. Unions are only used on cold pipe systems.
   b. Unions can be insulated by installing a larger section of pipe insulation over the union.
   c. Unions can be insulated with fiberglass blanket covered with insulation cement.
   d. Unions on cold applications are never insulated.

SUMMARY

The proper installation of pipe fittings, valves, and flanges is one of the most difficult tasks to master on your road to becoming an insulation mechanic. There is no shortcut to learning these skills, and you will spend a lot of time practicing in order to become accomplished.

The art of vapor sealing the insulation system at fittings must be mastered in cold service applications. The integrity of the system is only as good as the vapor retarder is applied at these fittings. A poorly sealed system can lead to damage to the system as well as injury to personnel.

The specifications for a project usually spell out the finish for fittings. This finish for fittings can be completely different than that for the straight run piping. Many times, the application of the specified finish is exceedingly difficult. For example, applying aluminum jacket to a flanged valve takes a considerable amount of skill. You must become proficient in applying all types of finish materials.

If you take the time to thoroughly master this module, the extra time spent in this area will prove to be very helpful as your career progresses.

1. Check valves are used for _____.
   a. reversing the direction of fluid flow
   b. alerting an operator to a high temperature
   c. mixing process materials
   d. limiting fluid flow to one direction

2. Flanges are generally _____.
   a. used on pipe of 2" or larger
   b. used to attach fitting covers
   c. used for 90-degree joints only
   d. glued in place

3. When insulating a valve body, you should _____.
   a. not cover the valve bonnet
   b. not cover the packing gland bolt heads
   c. not use scrap insulation for filling voids
   d. be sure the valve is on a hot pipe system.

4. In insulating, a 90-degree pipe turn is also known as a(n) _____.
   a. elbow
   b. return bend
   c. U-bend
   d. tee

5. A short-radius elbow has a radius equal to _____.
   a. the diameter of the insulation
   b. one-half the diameter of the pipe
   c. the diameter of the pipe
   d. one and one half times the diameter of the pipe

6. Reducers are used to reduce _____.
   a. the size of pipe in a run
   b. the temperature of material in a pipe
   c. the pressure of material in a pipe
   d. vibrations from hydraulic pressure

7. Which of the following can damage PVC fitting covers?
   a. Water hammers
   b. Low temperatures
   c. Ultraviolet light
   d. Misalignment

8. The outside turn of an ell is called the _____.
   a. end
   b. heel
   c. throat
   d. outside

9. The inside turn of an ell is called the _____.
   a. inside
   b. throat
   c. miter
   d. pitch

10. Before cutting miters on a bandsaw, what two things must be set?
    a. Time and date
    b. Pipe size and thickness
    c. Back stop and number of miters
    d. Heel and throat

11. When mitering a stove pipe fitting cover to go on a 45-degree bend, at what angle would you cut the insulation?
    a. $22\frac{1}{2}$ degrees
    b. 45 degrees
    c. $33\frac{1}{3}$ degrees
    d. 90 degrees

12. On small piping, a built-up fitting cover can be made from _____.
    a. fiberglass
    b. calcium silicate
    c. insulation cement
    d. loose fill

13. Before mixing cement for a cover, _____.
    a. check the temperature
    b. check the humidity
    c. measure the fitting
    d. read the SDS

14. Where should pipe insulation end when it is adjacent to a line flange?
    a. Butted tightly to the flange.
    b. A distance equal to the length of a section of insulation.
    c. 2" (5 cm) from the connection.
    d. A distance that will allow the removal of the flange bolts.

15. Tee insulation can be _____.
    a. cemented to the main pipe insulation
    b. sleeved over the main pipe insulation
    c. assembled from scrap
    d. built up using cement

Fill in the blank with the correct term that you learned from your study of this module.

1. The inside measure of an elbow or a mi-tered segment is a(n) _____.

2. A(n) _____ has a radius equal to one and one half times the pipe diameter.

3. The outside measure of an elbow or a mi-tered segment is a(n) _____.

4. _____ can be used to build up insula-tion on a fitting.

5. Segments of insulation that are cut to fit a curved fitting are called _____.

6. A(n) _____ has a radius equal to the pipe diameter.

7. The _____ of a circle is half its diam-eter.

8. To perform maintenance on a valve you would need to remove the _____.

9. A(n) _____ ensures a complete seal around a valve.

Trade Terms

Bonnet
Heel
Hydraulic setting insulation cement
Long-radius elbow
Miters

Packing gland
Radius
Short-radius elbow
Throat

Trade Terms Introduced in This Module

Bonnet: A cover for a valve body that can be removed to allow for valve maintenance and repair.

Heel: The inside curve of an elbow.

Hydraulic setting insulation cement: A type of quick-setting cement.

Long-radius elbow: A 45-degree or 90-degree elbow with its radius equal to one and a half times the diameter of the line.

Miters: Segments of insulation that fit together to insulate an elbow.

Packing gland: An assembly that houses a valve and contains a sealing material (packing), used to form a complete seal and prevent leakage.

Radius: The distance between the center point and the perimeter of a circle or arc.

Short-radius elbow: A 45-degree or 90-degree elbow with its radius equal to the diameter of the line.

Throat: The outside curve of an elbow.

Additional Resources

This module is intended as a thorough resource for task training. The following reference works are suggested for further study.

Pipe, Fittings, Valves, Supports, and Fasteners, International Pipe Trades Joint Training Committee. 2000. Annapolis, MD: United Association.

ASTM F683-10, Standard Practice for Selection and Application of Thermal Insulation for Piping and Machinery. ASTM International. Available at **www.astm.org/DATABASE.CART/HISTORICAL/F683-10.htm**

Figure Credits

Yury Maryunin/Hemera/Thinkstock, Module opener

Apollo Valves and Elkhart Products Corporation, Figure 1 (photo)

Watts Regulator Company, Figure 3

Mueller Streamline Co., Figure 6

Courtesy of Charlotte Pipe and Foundry, Figures 7, 8

Topaz Publications, Inc., Figure 9

Johns Manville, SA01

Answer	Section Reference	Objective
Section One		
1. c	1.1.0	1a
2. c	1.2.0	1b
3. b	1.3.0	1c
Section Two		
1. c	2.1.2	2a
2. a	2.2.4	2b
3. d	2.3.0	2c
4. b	2.4.4	2d

This page is intentionally left blank.

NCCER CURRICULA — USER UPDATE

NCCER makes every effort to keep its textbooks up-to-date and free of technical errors. We appreciate your help in this process. If you find an error, a typographical mistake, or an inaccuracy in NCCER's curricula, please fill out this form (or a photocopy), or complete the online form at **www.nccer.org/olf**. Be sure to include the exact module ID number, page number, a detailed description, and your recommended correction. Your input will be brought to the attention of the Authoring Team. Thank you for your assistance.

Instructors – If you have an idea for improving this textbook, or have found that additional materials were necessary to teach this module effectively, please let us know so that we may present your suggestions to the Authoring Team.

NCCER Product Development and Revision
13614 Progress Blvd., Alachua, FL 32615

Email: curriculum@nccer.org
Online: www.nccer.org/olf

❏ Trainee Guide   ❏ Lesson Plans   ❏ Exam   ❏ PowerPoints   Other _____

Craft / Level: _____   Copyright Date: _____

Module ID Number / Title: _____

Section Number(s): _____

Description: _____

_____

_____

_____

Recommended Correction: _____

_____

_____

_____

_____

Your Name: _____

Address: _____

_____

Email: _____   Phone: _____

This page is intentionally left blank.

Glossary

All-service jacketing (ASJ): A facing or covering applied to piping and mechanical equipment fiberglass insulation as a vapor barrier or for protection against abrasion.

Ambient: Relating to the immediate surroundings of an object, system, or person.

Architect: A person whose profession is to design and create plans for buildings, bridges, and facilities.

Backflow preventer: A device that prevents liquid from flowing backward in the event of a loss in pressure.

Boilers: Pressure vessels in which water is turned into steam for heating purposes, for operating equipment, or to generate power.

Bonnet: A cover for a valve body that can be removed to allow for valve maintenance and repair.

Booster pump: A pump that increases the pressure of a liquid to provide adequate pressure for the upper floors in a facility.

Branch: A part of a distribution piping system that is usually smaller than a main and used to connect a main to two or more runouts.

Breechings: The ducts or pipes connecting the exhaust-gas discharge from a boiler furnace to the stack.

Butt strips: Strips of ASJ that are several inches wide used to secure ends of piping insulation ASJ to each other after installation.

Butt weld: A pipe connection made by beveling the ends of two pieces of pipe and welding them together.

Chase: A vertical enclosure that houses piping and ductwork in a structure.

Chilled water: Water at below-ambient temperature used for cooling (particularly in air conditioning systems or in processes). Typical chilled water temperature for comfort cooling applications is 42°F–55°F (7°C–13°C).

Circumferential joint: A break in the insulation encircling the pipe, formed when two sections of insulation are butted together.

Condensation: Water formed when ambient air comes in contact with something cooler; The physical process by which a liquid is removed from its vapor by cooling (e.g. water vapor turns into a liquid upon contact with a cold surface).

Conditioned air: Air that has been treated to control its temperature, humidity, and/or cleanliness to meet the requirements of a conditioned space.

Construction elevator: An elevator dedicated to the movement of material and personnel.

Copper tubing size (CTS): A pipe sizing standard. CTS pipe is characterized by thinner walls than iron pipe size.

Crew: Mechanics and helpers assigned to a project. Each craft will have a crew for their specific applications.

Diameter: The distance across a circle measured through its center.

Distribution system: A piping system consisting of mains, risers, drops, branches, tanks, pumps, valves, fixture connections, and additional equipment.

General contractor: A firm that usually manages all crafts involved in constructing a facility.

Heat tracing: A form of applying heat to the outside surface of a pipe to keep the internal product at a certain temperature.

Heating, ventilating, and air conditioning (HVAC): A type of system designed to provide thermal comfort and acceptable air quality. The majority of HVAC systems provide both cooling and heating, although some provide only cooling and others provide only air heating.

Heel: The inside curve of an elbow.

HMIS label: A safety label required by the Federal Hazardous Materials Information System (HMIS) that is applied to all containers holding hazardous materials used in construction and industrial, government, and commercial activities.

Hot water heating: Water heated by natural or artificial means used for structural comfort heating or in processes. Typical hot water temperature for comfort heating applications is 170°F–190°F (77°C–88°C).

Hydraulic setting insulation cement: A type of quick setting cement.

Industrial plant: A facility that produces electricity or other products used by the general population, such as gasoline and plastics.

Inside diameter (ID): The inside measurement of a section of insulation and/or the inside measurement of a section of pipe.

Iron pipe size (IPS): A pipe sizing standard. IPS is characterized by thicker walls than copper tubing size.

Long-radius elbow: A 45-degree or 90-degree elbow with its radius equal to one and a half times the diameter of the line.

Longitudinal joint: A break in the insulation formed by the split provided in molded fiberglass pipe insulation along its length on one side. The longitudinal joint is opened to slide the insulation over a pipe, then sealed after installation.

Main: The principal distribution piping that connects a supply source to all of the branches in a system.

Material storage area: A space located at the job site where all materials are stored.

Mechanic: An individual with deep knowledge of the applications of a specific mechanical craft.

Mechanical contractor: A firm that installs systems in commercial/industrial facilities. Some mechanical contractors are trade-specific, some do plumbing and HVAC, some plumbing only, and some HVAC only.

Mechanical insulation: The thermal, acoustical, and personnel safety systems applied to piping, ductwork, and equipment.

Mechanical insulator: A craftworker who applies and installs mechanical insulation systems.

Miters: Segments of insulation that fit together to insulate an elbow.

Nesting: Taking different sizes of like insulation and placing the smaller size inside the larger size to make the overall thickness greater.

NFPA fire diamond: A safety label established by the National Fire Protection Association (NFPA) used by emergency responders to understand the hazards of chemicals involved in a fire.

Nominal: The target for a given measurement during manufacturing, such as the thickness of insulation. The material may actually be slightly less than or slightly greater than the nominal measurement.

Non-potable water: Water that is not suitable for drinking, such as graywater / reclaimed water.

Occupational Safety and Health Administration (OSHA): The federal government agency established to ensure a safe and healthy environment in the workplace.

On-the-job learning (OJL): Job-related learning an apprentice acquires while working under the supervision of journey-level workers. Also called *on-the-job training (OJT)*.

Outside diameter (OD): The outside measurement of a section of insulation or a section of pipe.

Packing gland: An assembly that houses a valve and contains a sealing materials (packing), used to form a complete seal and prevent leakage.

Packing slip: A detailed listing of all materials shipped from a supplier. Every delivery/shipment has a packing slip.

Pallets: Small platforms used to place boxes on, to keep materials off the ground, and to make moving and delivering bulk materials easier.

Personnel protection: Insulation applied only where employees may come in contact with hot pipes or equipment.

Pipe insulation thickness: The thickness of insulation to be applied to the pipe system.

Pipe saddle: A pipe support formed from metal curved to conform to the shape of the pipe.

Pipe shoe: A pipe support made of tee iron and welded to the bottom of the pipe.

Pipe size: The measurement, expressed in inches, of the outside diameter (OD) and inside diameter (ID) of the pipe.

Pipe systems: Sections of pipe connected together to allow the movement of different processes.

Plenum: An enclosed space in a facility, often located above a workspace between the structural ceiling and a drop-down ceiling, which allows air circulation for HVAC systems.

Plumbing system: A system of pipes, valves, and fittings that carries water to and drains wastewater from a building, facility, or home.

Polyethylene sheeting: For storage purposes, heavy flexible sheets of plastic used as tarpaulins for protecting materials from the elements.

Potable water: Water that is suitable for drinking (for example, domestic water, city water, or well water).

Puncture: A small hole formed in an otherwise continuous material. In insulating, it tends to apply to holes made in jacket during installation.

Radius: The distance between the center point and the perimeter of a circle or arc.

Recirculating hot water: A system in which hot water is pumped in a continuous cycle to provide instant hot water at a fixture.

Refrigeration: The process of transferring heat from one substance or area to another in order to lower the temperature of the substance or area.

Riser: Vertical piping to or from a main, branch, or runout.

Rope and pulley: A hoisting system consisting of a rope running around grooved wheels, which makes lifting heavy loads easier.

Runout: Piping to or from a branch or main to a plumbing unit or fixture connection, a heating and/or cooling unit connection, or a process equipment connection.

Scaffolds: Elevated work platforms for both personnel and materials.

Self-sealing lap system: An adhesive-backed strip that seals the ASJ lap (longitudinal) joint.

Short-radius elbow: A 45-degree or 90-degree elbow with its radius equal to the diameter of the line.

Socket weld: A pipe connection made by sliding a collar over the two pieces of pipe to be joined and then welding the collar to the pipes.

Stack: An orderly pile of like materials.

Staging: The placement of materials in smaller stacks close to the actual work site for easy access.

Tarpaulin: Waterproofed fabric or canvas material that is used to cover insulation materials.

Throat: The outside curve of an elbow.

Trapeze hangers: Pipe hangers with parallel vertical rods that are suspended from a structure and connected at the bottom with a horizontal member from which one or more pipes can be supported.

Vapor retarder: A material designed to minimize the passage of water vapor between two volumes of air. Vapor retarders are sometimes (incorrectly) referred to as vapor barriers, but all vapor retarders are permeable to vapor to some extent.

Water meter: A device that measures and records the amount of water that passes through a pipe or similar object.

Water softeners: Pieces of equipment that remove minerals from water.

This page is intentionally left blank.

Index

Material storage area, (19104):1, 11
Mechanical contractor, (19101):1, 7, 26
Mechanical engineers, (19210):1
Mechanical insulation, (19101):1, 26
Mechanical insulation mechanic, (19101):1
Mechanical insulators
  apprenticeship programs, (19101):6–7, 15
  career paths, (19101):6–7
  defined, (19101):1, 26
  employment, (19101):7
  good, qualities of, (19101):8
  PPE requirements, (19209):2
  tool maintenance and storage, (19101):13
  training, (19101):6–7, 15
  work done by, (19101):1
  working conditions, (19101):7–8
Mechanics. *See also* Mechanical insulators
  defined, (19101):1, 26
  mechanical insulation mechanics, (19101):1
  PPE requirements, (19209):2
Melamine foam insulation, (19101):2
Microporous insulation, (19101):3
Mineral fiber/mineral wool insulation, (19101):2
Miters, (19107):4, 9–13, 23
Mixing and diverting valves, (19107):1
Molded fiberglass insulation, (19106):1

N
NASA, (19101):5
NCCER
  Registered Apprenticeship program for mechanical
    insulation, (19101):16
  Registry System, (19101):16
  training credentials, (19101):27
  training curricula, (19101):16
Needles, hex mesh wire, (19101):11
Nesting, (19105):1, 19
NFPA fire diamond, (19104):5, 7, 11
90-degree ells, (19107):4
Nippers, end-cutting, (19101):10
No-hub cast iron pipe, (19105):6, 7
Nominal, (19105):1, 19
Non-potable water, (19209):1, 3, 8, 9–10, 14

O
Obedience, expectations for, (19101):19
Occupational Safety and Health Administration (OSHA)
  defined, (19101):18, 26
  employer responsibilities under, (19101):20
  reporting requirements, (19101):20
  Safety and Health Standards for the Construction Industry,
    (19101):20
OD. *See* Outside diameter (OD)
OJL. *See* On-the-job (OJL) learning
Open cell insulation, (19101):1–2
Orbiter Space Shuttle, (19101):5
OSHA. *See* Occupational Safety and Health Administration
  (OSHA)
Outside diameter (OD), (19105):1, 19

P
Packing gland, (19107):1, 2, 23
Packing slip, (19104):1, 2, 11
Pallets, (19104):1, 2, 11
Perlite insulation, (19101):3
Personal protective equipment (PPE), (19101):6, (19104):5,
  (19209):2
Personnel protection, (19101):1, 26
Phenolic insulation, (19101):2

Pipe. *See also specific types of*
  construction
    cast iron pipe, (19105):5, 7
    copper pipe and tubing, (19105):5–6, 7
    plastic pipe, (19105):6–8
    steel pipe, (19105):4–5
  recycling, (19209):2
  thermal expansion by type of pipe, (19105):13
Pipe bosses, (19106):6
Pipefitter, (19107):4
Pipe fittings
  cast iron, (19107):6
  connections
    cemented (glued) joints, (19107):5
    mechanical couplings, (19107):5
    soldered and brazed joints, (19107):5
    threaded joints, (19107):4, 5, 6
    welded joints, (19107):5
  copper, (19107):5
  function, (19107):4
  insulation
    blanket, (19107):13–14
    built-up covers, (19107):13–14
    cements, (19107):14
    mitered segments, (19107):9–13
    molded or machined, (19107):9
    PVC covers, (19107):6–9
  plastic, (19107):6
  steel, (19107):5–6
  types of
    45-degree ells, (19107):4
    90-degree ells, (19107):4
    caps, (19107):4
    reducers, (19107):4
    return bends, (19107):4
    tees, (19107):4
Pipe insulation
  chilled water systems, (19210):8
  hangers sized for, (19105):9
  return air plenums, (19209):10
  sizing, (19106):1
  thickness, (19105):1, 2–3, 19, (19106):1, 2, 14
Pipe saddle, (19106):4, 8, 14
Pipe shoe, (19106):4, 8, 14
Pipe size
  copper tubing (CTS), (19105):1–2, 3–4, 19
  defined, (19105):1, 19
  inside diameter (ID), (19105):1, 19
  insulation thickness, (19105):1, 2–3, 19
  interchangeable, (19105):3
  iron pipe (IPS), (19105):1–2, 3–4, 17, 19
  length, (19105):1
  outside diameter (OD), (19105):1, 19
  wall thickness, (19105):1
Pipe systems
  cold water piping
    components, (19209):4–5
    hangers and supports, (19105):9, 10
    materials, (19209):2
    system layout, (19209):3–4
  components
    anchors, (19105):9, 11
    earthquake systems, (19105):11
    expansion loops and joints, (19105):9, 13
    hangers, (19105):9, 10
    heat tracing systems, (19105):9, 11–12, 13, 19
    supports, (19105):9
    valves, (19105):12, 14

condensate drains, (19209):9
defined, (19105):1, 19
hot water piping
   hangers and supports, (19105):9, 11
   materials, (19209):6
   system layout, (19209):7–8
reclaimed water, (19209):9–10
return air plenums, (19209):10
storm drains, (19209):8–9
Pipe vs. tube, (19105):8
PIR. *See* Polyisocyanurate (PIR) insulation
Pistol grip bander, (19101):10, 11
Plastic pipe
   applications, (19105):6–7
   cemented (glued) joints, (19107):5
   condensate drain piping, (19209):9
   construction, (19105):6–8
   fittings, (19107):6
   Pipe and Insulation Chart, (19105):18
   reclaimed water piping systems, (19209):10
   return air plenums piping systems, (19209):10
Plenum, (19209):1, 8, 14
Plumbing system, (19101):1, 7, 26
Polyethylene sheeting, (19104):1, 2, 11
Polyimide foam insulation, (19101):2
Polyisocyanurate (PIR) insulation, (19101):2
Polyolefin insulation, (19101):2
Polyvinyl chloride (PVC) fitting covers, (19107):6–9
Polyvinyl chloride (PVC) pipe
   advantages/disadvantages, (19209):2
   construction, (19105):6–8
   fitting insulation, (19107):6–9
   Pipe and Insulation Chart, (19105):18
Potable water, (19209):1, 2, 14
PPE. *See* Personal protective equipment (PPE)
Professionalism, expectations of, (19101):18
P-traps, (19209):9
Puncture, (19106):1, 14
PVC. *See* Polyvinyl chloride (PVC) fitting covers; Polyvinyl chloride (PVC) pipe

R

Radius, (19107):4, 23
Ratchet bander, (19101):11
Recirculating hot water, (19209):1, 8, 14
Reclaimed water piping systems, (19209):9–10
Reducer, (19107):4
Reducer covers, (19107):17
Reflective insulation, (19101):3, 4, 5
Refractory insulation, (19101):3
Refrigerants, (19210):1
Refrigeration, (19209):1, 9, 14
Relief valves, (19107):1, 3
Respectfulness, expectations for, (19101):19
Responsibility, taking, (19101):19
Return air plenums piping systems, (19209):10
Return bends, (19107):4
Rigid board insulation, (19101):5
Rigid foam insulation, (19101):2
Riser, (19209):1, 4, 14
Roller, sheet metal, (19101):12, 13
Roof drains, (19209):9
Rope and pulley, (19104):5, 11
Rules and regulations, expectations for following, (19101):19
Runout, (19209):1, 4, 14

S

Saddle cutting insulation, (19107):16

Safety
   cements, working with, (19107):14
   employee responsibility for, (19101):19–20
   hoisting, (19104):6
   insulation, (19101):5–6
   moving materials, (19104):5, 6–7
Safety data sheet (SDS), (19101):6, (19107):14, 15
Safety labels, (19104):7–8
Safety training, (19101):20
Scaffolds, (19104):1, 3, 11
Schedulers, (19101):7
SDS. *See* Safety data sheet (SDS)
Self-sealing lap system, (19106):1, 2, 14
Semi-rigid board insulation, (19101):5
Shears
   sheet metal, (19101):10, 11
   sheet metal jacketing, (19101):12, 13
Sheet metal, storing, (19104):4
Sheet metal shears, (19101):10, 11, 12, 13
Short-radius elbow, (19107):4, 23
Site superintendents, (19101):7
Socket weld, (19105):1, 5, 6, 19
Solar power hot water systems, (19209):7
Soldering
   copper tubing, (19105):5
   joints, (19107):5
Spray-on fiber, granular, cement materials insulation, (19101):5
Stack, (19104):1, 2, 11
Staging, (19104):5, 6, 11
Stainless steel pipe, (19105):4, 17
Steam diversion systems, (19210):10
Steel pipe
   applications, (19105):4
   construction, (19105):4–5
   fittings, (19107):5–6
   thermal expansion, (19105):13
Storm drains piping systems, (19209):8–9
Stove pipe miters, (19107):12–13
Structural insulation, (19101):6
Superintendents, (19101):7
Supervisors, (19101):7
Supports, pipe, (19105):9

T

Tankless hot water heaters, (19209):8
Tape insulation, (19101):5
Tardiness, (19101):19
Tarpaulin, (19104):1, 2, 11
Tees, (19107):4, 16
Temperature, insulation categorized by, (19101):3, 4
Textile-based fibrous insulations, (19101):2
Textile glass insulation, (19101):2
Thermal insulating coatings, (19101):3
Threaded joints, (19107):4
Thredolets®, (19106):6
Throat, (19107):4, 9, 23
Training
   credentials, NCCER, (19101):27
   curricula, NCCER, (19101):16
   insulation mechanics, (19101):6–7, 15
   safety training, (19101):20
Trapeze hangers, (19209):1, 4, 14
Trowels
   plastering, (19101):10, 11
   pointing, (19101):10
Tube vs. pipe, (19105):8

U

Unions, (19107):17

V

Valves
  insulation installation on
    flanged valves, (19107):2
    non-flanged valves, (19107):2
  types of
    balancing, (19107):1
    butterfly, (19107):1
    check, (19107):1, 3
    control, (19107):1
    gate, (19107):1, 2
    mixing and diverting, (19107):1
    relief, (19107):1, 3
Vapor retarder, (19106):1, 2, 14, (19209):5, 8
Vermiculite, (19101):3

W

Waste steam, (19210):10
Wastewater, (19209):9–10
Water
  chilled, (19210):1, 16
  non-potable, (19209):1, 8, 9–10, 14
  potable water, (19209):1, 14
  reclaimed/recycled, (19209):8, 9–10
Water hammer arrestors, (19209):5, 6
Water heaters, (19209):7–8
Water meter, (19209):1, 3, 14
Water softeners, (19209):1, 14
Welding pins, storing, (19104):4
Weldolets®, (19106):6
Welds
  butt weld, (19105):1, 4, 6, 19
  socket weld, (19105):1, 5, 6, 19
  steel pipe, (19105):4–5
  welded joints, (19107):5
Wool-based fibrous insulations, (19101):2
Workflow, new construction projects, (19101):8
Wrought iron pipe, thermal expansion, (19105):13

X

XPS. *See* Extruded polystyrene foam (XPS) insulation

www.ingramcontent.com/pod-product-compliance
Lightning Source LLC
Chambersburg PA
CBHW081555220326
41598CB00036B/6686